Fiber-Optic Fabry–Perot Sensors

Series in Fiber Optic Sensors
Alexis Méndez, Series Editor

Now Available

An Introduction to Distributed Fibre Sensors
Arthur H. Hartog

Forthcoming Titles

Fiber Optic Structural Health Monitoring (SHM) and Smart Structures
Alfredo Güemes and Julian Sierra Perez

Introduction to Fiber Bragg Grating Sensors
John Canning, Cicero Martelli, and Kyriacos Kalli

Fiber-Optic Fabry–Perot Sensors

An Introduction

Yun-Jiang Rao
Zeng-Ling Ran
Yuan Gong

CRC Press
Taylor & Francis Group
Boca Raton London New York

CRC Press is an imprint of the
Taylor & Francis Group, an **informa** business

CRC Press
Taylor & Francis Group
6000 Broken Sound Parkway NW, Suite 300
Boca Raton, FL 33487-2742

CRC Press is an imprint of Taylor & Francis Group, an Informa business

International Standard Book Number-13: 978-1-1383-0579-3 (Paperback)

Library of Congress Cataloging-in-Publication Data

Names: Rao, Yun-Jiang, author. | Ran, Zeng-Ling, author. | Gong, Yuan (Associate professor), author.
Title: Fiber-optic Fabry-Perot sensors : an introduction / Yun-Jiang Rao, Zeng-Ling Ran, Yuan Gong.
Description: Boca Raton, FL : CRC Press, Taylor & Francis Group, [2017] | Includes bibliographical references and index.
Identifiers: LCCN 2016049401| ISBN 9781498736930 (hardback ; alk. paper) | ISBN 1498736939 (hardback ; alk. paper)
Subjects: LCSH: Fabry-Perot interferometers. | Optical fiber detectors.
Classification: LCC QC411.4 .R36 2017 | DDC 681/.25--dc23
LC record available at https://lccn.loc.gov/2016049401

Visit the Taylor & Francis Web site at
http://www.taylorandfrancis.com

and the CRC Press Web site at
http://www.crcpress.com

Contents

Series Preface

Optical fibers are considered among the top innovations of the twentieth century, and Sir Charles Kao, a visionary proponent who championed their use as a medium for communication, received the 2009 Nobel Prize in Physics. Optical fiber communications have become an essential backbone of today's digital world and Internet infrastructure, making it possible to transmit vast amounts of data over long distances with high integrity and low loss. In effect, most of the world's data flows nowadays as light photons in a global mesh of optical fiber conduits. As the optical fiber industry turns 50 in 2016, the field might be middle aged, but many more advances and societal benefits are expected of it.

What has made optical fibers and fiber-based telecommunications so effective and pervasive in the modern world? It is its intrinsic features and capabilities make it so versatile and very powerful as an enabling and transformative technology. Among its characteristics we have electromagnetic (EM) immunity, intrinsic safety, small size and weight, capability to perform multipoint and multiparameter sensing remotely, and so on. Optical fiber sensors stem from these same characteristics. Initially, fiber sensors were lab curiosities and simple proof-of-concept demonstrations. Nowadays, however, optical fiber sensors are making an impact and serious commercial inroads in industrial sensing, biomedical applications, and military and defense systems, and have spanned applications as diverse as oil well downhole pressure sensors to intra-aortic catheters.

This transition has taken the better part of 30 years and has now reached the point where fiber sensor operation and instrumentation are well developed and understood, and a variety of commercial sensors and instruments are readily available. However, fiber sensor technology is not as widely known or deeply understood today as other more conventional sensors and sensing technologies such as electronic, piezoelectric, and MEMS devices. In part, this is due to the broad range of different types of fiber sensors and techniques available. On the other hand, although there are several excellent textbooks reviewing optical fiber sensors, their coverage tends to be limited and they do not provide sufficiently in-depth reviews of each sensor technology type. Our book series aims to remedy this by providing a collection of individual tomes, each focused exclusively on a specific type of optical fiber sensor.

The goal of this series has been from the onset to develop a set of titles that feature an important type of sensor, offering up-to-date advances as well as practical and concise information. The series encompasses the most relevant and popular fiber sensor types in common use in the field, including fiber Bragg grating sensors, Fabry–Perot sensors, interferometric sensors, distributed fiber sensors, polarimetric sensors, polymer fiber sensors, structural health monitoring (SHM) using fiber sensors, biomedical fiber sensors, and several others.

This series is directed at a broad readership of scientists, engineers, technicians, and students involved in relevant areas of research and study of fiber sensors, specialty optical fibers, instrumentation, optics, and photonics. Together, these titles will fill the need for concise, widely accessible introductory overviews of the core technologies, fundamental design principles, and challenges to the implementation of optical fiber-based sensors and sensing techniques.

This series has been made possible due to the tenacity and enthusiasm of the series manager, Ms. Luna Han, to whom I owe a debt of gratitude for her passion, encouragement, and strong support—from the initial formulation of the book project and throughout the full series development. Luna has been a tremendous resource and facilitator, and a delight to work with in the various stages of development for each of the series volumes.

Information, as the saying goes, is knowledge. And thanks to the dedication and hard work of the individual volume authors as well as chapter coauthors, the readers have enriched their knowledge on the subject of fiber-optic sensors. I thank all the authors and extend my deep appreciation for their interest and support for this series and for all the time and effort they poured into its writing.

To the readers, I hope that this series is informative, fresh, and of aid in their ongoing research, and wish them much enjoyment and success!

Alexis Méndez, PhD
Series Editor
President
MCH Engineering, LLC
Alameda, CA

Preface

Optical fiber sensors (OFS) are of great importance in the sensor community, as they have a number of advantages over conventional sensors, such as immunity to electromagnetic interference, capability of responding to a wide variety of measurands, very high resolution, high accuracy, small size, long-distance, ability to operate in harsh environments, etc. OFS mainly consist of two categories, that is, point OFS and distributed OFS, which have been widely used for health and safety monitoring in many fields. Among the point OFS family, the fiber-optic Fabry–Perot interferometric (FFPI) sensor is one of the most important members. The FFPI was proposed and demonstrated about 35 years ago. To date, various FFPIs, such as strain, pressure, temperature, and vibration sensors, have been successfully commercialized. From the point of view of practical applications, conventional FFPIs with handmade cavities suffer from the problems of reproducibility, repeatability, and long-term stability under harsh environments. From a scientific research point of view, conventional FFPIs based on single-mode or multimode fibers have been studied extensively, and it is difficult to advance FFPIs further.

About 10 years ago, the use of advanced fabrication techniques based on laser micromachining paved the way to realize novel in-line all-fiber FFPIs, which have many advantages over conventional

FFPIs, for example, direct formation, mass production, low cost, good reproducibility, excellent stability, and high-temperature ability, making it ideal to form a new generation of FFPIs for various practical applications. In particular, the FFPI strain sensor is considered to be the best high-temperature strain sensor reported so far and has been successfully applied to high-temperature strain measurements of ultrahigh-speed aero-vehicles and aero-engines, and has broken the bottleneck of traditional strain gauges in both accuracy and size, as well as solved the well-known problem of precise strain measurement under high temperature. Also, novel FFPIs based on photonic crystal fibers were proposed and demonstrated, providing new ways for temperature-insensitive and high-precision measurement of the refractive index, which would find important applications in chemical and biomedical sensing. Such work has attracted great interest from the OFS society. These novel FFPIs can provide revolutionary means for optically precise measurements of a wide range of parameters under harsh environments.

This book gives a comprehensive overview of the progress in FFPIs over the last 35 years and also includes many recent research results achieved by both the authors and the whole community. There are six chapters covering the following areas: Principles; Structures; Fabrication; Physical and Biochemical Sensors; Interrogation and Multiplexing; and Applications.

I thank the members of the Fiber Optics Research Center of the Key Laboratory of Optical Fiber Sensing & Communications (Ministry of Education) at the School of Communication & Information Engineering, University of Electronic Science & Technology of China, Chengdu, Sichuan, China, for their significant contributions over the last 10 years. I also thank the members of the Optical Fiber Technology Group, which I established in 1999 at Chongqing University, Chongqing, China, for their contributions during 2000–2010.

Also, I thank my wife, Huang Juan, and my son, Rao Oufan, for their unconditional and continuous support. Finally, I like to thank Ms. Luna Han for her support as the production editor of the OFS book series at Taylor & Francis Group.

It is our hope that this book will provide a good reference for those researchers and engineers who are interested in the FFPI technique and its applications.

Professor Yun-Jiang Rao
IEEE Fellow, OSA Fellow, SPIE Fellow
Founding Editor-in-Chief, Photonic Sensors (ESCI, Springer)
Director, Fiber Optics Research Center
Key Lab of Optical Fiber Sensing & Communications (MOE)
University of Electronic Science and Technology of China
Chengdu, Sichuan, China

Authors

Yun-Jiang Rao was dean of the School of Communication & Information Engineering and the School of Graduates at the University of Electronic Science and Technology of China (UESTC) during 2005–2015, and is currently the Vice Chairman, Academic Committee of UESTC and the director of the Key Lab of Optical Fiber Sensing & Communications of Ministry of Education in China. Prior to working there, he was a Postdoctoral Research Fellow at the University of Strathclyde in Scotland, UK, and then a Research Fellow/Senior Research Fellow at the University of Kent in Canterbury, UK. He has been working on fiber-optic Fabry–Perot and grating sensors since he joined the groups of two outstanding pioneers of fiber-optic sensors, Professor Brian Culshaw and Professor David Jackson, in 1990s. He has authored or coauthored more than 400 peer-reviewed papers in scientific journals and international conference proceedings, which gained ~8000 citations in Google Scholar (H-index of 43). He serves as an associate editor of *IEEE/OSA Journal of Lightwave Technology*, and is also the founding editor-in-chief of *Photonic Sensors (Springer)*. He founded the Asia-Pacific Optical Sensor Conferences (APOS) and has been serving as a TPC member for the OFS conferences since 2008. He is a fellow of the IEEE, OSA and SPIE.

Zeng-Ling Ran is a professor at the Fiber Optics Research Center, Key Lab of Optical Fiber Sensing & Communications (Ministry of Education) at the University of Electronic Science and Technology of China (UESTC).

Yuan Gong is an associate professor at the Fiber Optics Research Center, Key Lab of Optical Fiber Sensing & Communications (Ministry of Education) at the University of Electronic Science and Technology of China (UESTC).

1 PRINCIPLES

1.1 Physics of Fabry–Perot Cavities

A Fabry–Perot (FP) is an optical cavity device that acts as an interferometer and thus, it is also known as a Fabry–Perot interferometer (FPI). A FPI is constructed of two optical reflectors M_1 and M_2 with reflectances R_1 and R_2 on either side of an optically transparent medium with distance of length h, as shown in Figure 1.1. The theory of FPI is reviewed in detail by Born and Wolf [1] as well as by Hernandez [2]. Charles Fabry and Alfred Perot invented FPI in 1899 [3]. Since then, FPI has been widely used for high-resolution spectroscopy. In 1981, Douglas L. Franzen and Ernest M. Kim first connected optical fibers to FP geometry [4]. The basic idea of using a single fiber to realize a fiber-optic FP cavity was presented by Cielo [5] and by Yoshino and Ohno [6]. Petuchowski et al. [7] discussed the implementation of a singlemode fiber-optic FPI (FFPI) with fiber ends to serve as mirrors. On account of the advantages of FFPI cavity, such as immunity to electromagnetic interference, capability of responding to a wide variety of parameters, very high resolution, high accuracy, and small size, it becomes an ideal transducer for many sensing applications. Some of the FFPI sensors have been successfully commercialized and widely used for health monitoring of composite materials, large civil engineering structures (e.g., bridges and dams), space aircrafts, airplanes, etc., which would lead to the realization of the so-called smart materials and structures [8].

As an interferometer, we assume transmittance T_i and reflectance R_i, $i = 1, 2$, such that $R_i + T_i = 1$. We neglect the excess loss, relying on the absorption and scattering. The reflectance R and transmittance T are calculated using the following equations [9]:

$$R = \frac{R_1 + R_2 + 2\sqrt{R_1R_2}\cos\phi}{1 + R_1R_2 + 2\sqrt{R_1R_2}\cos\phi} \tag{1.1}$$

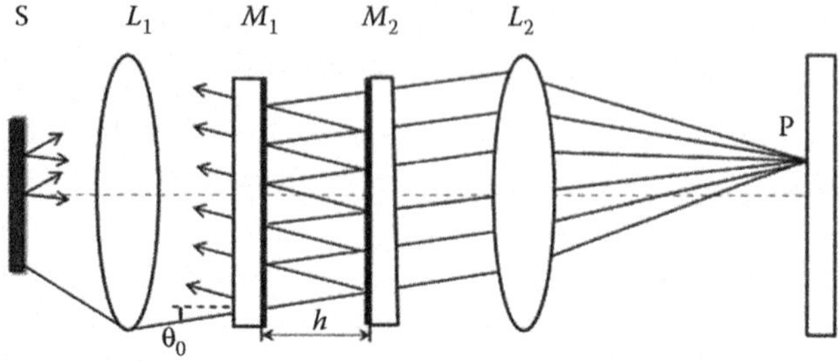

Figure 1.1 Schematic of FP cavity.

$$T = \frac{T_1 T_2}{1 + R_1 R_2 + 2\sqrt{R_1 R_2}\cos\phi} \tag{1.2}$$

where R is the ratio of the intensity reflected by FPI I_r, to the incident intensity I_i, T is the ratio of the transmitted intensity I_t to the incident intensity, and ϕ is the propagation phase shift in the interferometer, which is calculated by

$$\varphi = \frac{2\pi}{\lambda}\Delta \tag{1.3}$$

where Δ is the optical path difference (OPD), mainly relying on the characteristics of the cavity and λ is the free-space optical wavelength.

Equation 1.2 shows that T reaches a maximum for $\cos\phi = -1$ equal to $\phi = (2m + 1)\pi$, when m is integral. So, the intensity of the optical spectrum changes periodically along with the optical wavelength for a certain cavity. Defining $\delta = \varphi - (2m + 1)\pi$, $\cos\phi \approx -(1 - \delta^2/2)$, then T approaches a maximum, with $\delta \ll 1$. If the reflectivities of the optical mirrors are equal and approach unity, Equation 1.2 can be simplified to

$$T = \frac{T^2}{(1-R)^2 + R\delta^2} \tag{1.4}$$

where $R = R_1 = R_2$ and $T = T_1 = T_2 = 1 - R$. When $\delta = 0$, the transmittance reaches to the maximum. The half-high width, $\Delta\varphi$, is defined as the phase difference between the transmittance peak and the adjacent half-maximum point. Equation 1.4 shows that, when δ is 0, T is the maximum and T is half of its maximum value for $\delta = \pm(1-R)/\sqrt{R}$. So, $\Delta\varphi$ can be calculated by

$$\Delta\varphi = \frac{2(1-R)}{\sqrt{R}} \tag{1.5}$$

The fineness N, a frequently used figure of merit for FP cavity, is defined as the ratio of the phase shift between adjacent transmittance peaks, 2π to $\Delta\varphi$, which can be written as

$$N = \frac{\pi\sqrt{R}}{(1-R)} \tag{1.6}$$

Aiming at the filling material between the two flats of the cavity, the FFPI sensors are generally classified into intrinsic and extrinsic ones. No matter what kind the sensor is, the fiber plays an important role in FFPI. Light emitted from the light source propagates along the fiber to the interference structure. Then, the detected signal couples to the photodetector along the fiber. The differences between intrinsic and extrinsic FFPI sensors mainly focus at the filling material of the cavity. For intrinsic FFPI sensors, the cavity formed by two separated mirrors is filled by the fiber itself, as shown in Figure 1.2a. But for extrinsic FFPI sensors, the filling of the cavity can be an air gap or some other materials except for the fiber, as shown in Figure 1.2b.

However, in this book, we will classify the FFPI sensors in another way, by focusing on the quantity of the interference beams. In the following sections of this chapter, we will introduce three major types of FFPI.

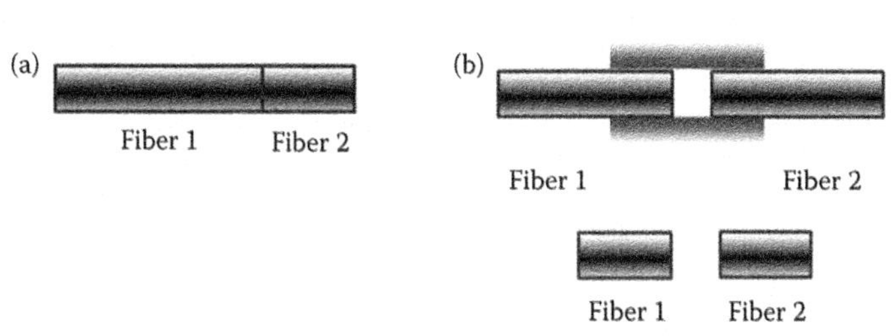

Figure 1.2 Structures of (a) intrinsic and (b) extrinsic FFPI cavities.

1.2 Principles of Fiber-Optic Fabry–Perot Interferometric (FFPI) Sensors

The history of FP sensors began at the turn of the nineteenth century with derivatives of the parallel-plate interferometer. Sensors for voltage and pressure measurement were described by Perot and Fabry [10]. After that, Meggers and Peters used FP interference to measure the refractive index [11]. In this book, we focus on FFPI sensors which are based on two-beam, three-beam, and multi-beam interferences, respectively.

The principles of FFPI sensors are mainly explained according to Equation 1.3. When perturbation is introduced to the sensor, the phase difference is influenced with the variation in the OPD of the interferometer. By measuring the shift of the phase or wavelength spectrum, the sensing parameter applied on the FFPI sensor can be quantitatively obtained. The free spectral range (FSR), the spacing between adjacent interference peaks in a spectrum, is also influenced by the OPD variation.

1.2.1 Two-Beam Interference

The principle of two-beam interference is the coherent addition of two reflectors with low reflectance. The typical structure is formed by the Fresnel reflections of two fiber ends. As shown in Figure 1.3a, supposing the intensity of the incident light is I, the reflectance of the fiber end is r, that is, ~4%. The transmission t is ~96%. I_{nr} and I_{nt} are the reflective intensity and transmitted intensity of the nth reflected beam, respectively. The intensities of the light are shown in Table 1.1. The intensities from the reflected end are I_{1r}, I_{3t}, I_{5t}, and so on.

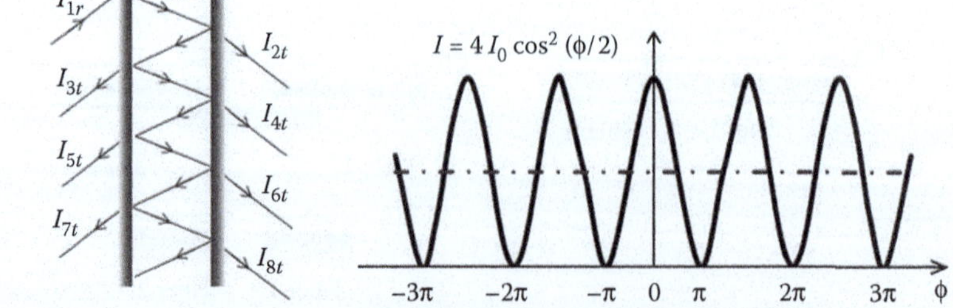

Figure 1.3 Two-beam interference: (a) schematic and (b) intensity distribution.

Table 1.1 Reflective and Transmitted Intensities from a FPI with Low Fresnel Reflection of 4%

INTENSITY	EQUATION	VALUE
I_{1r}	$r \cdot I$	4% I
I_{2t}	$t^2 \cdot I$	92.16% I
I_{3t}	$t^2 r \cdot I$	3.69% I
I_{4t}	$t^2 r^2 \cdot I$	0.15% I
I_{5t}	$t^2 r^3 \cdot I$	0.006% I
I_{6t}	$t^2 r^4 \cdot I$	2.4×10^{-6} I

Obviously, after the third reflection, the intensity I_{3r} is much lower than the first two, so the influence of I_{3r} can be neglected. It means that it is enough to consider only two reflected beams.

Suppose that the length of the cavity is h, and the refractive indexes of the fiber and the medium between the reflectors are n and n_0, respectively. The OPD is $\Delta = 2nh + \lambda/2$, where $\lambda/2$ is the half-wave loss because the two beams of the interference occurred by two planes with different properties.

The intensity of the interference, I, can be calculated as

$$I = I_1 + I_2 + 2\sqrt{I_1 I_2}\cos\frac{2\pi\Delta}{\lambda} \tag{1.7}$$

$$V = \frac{I_M - I_m}{I_M + I_m} = \frac{2\sqrt{I_2/I_1}}{1 + I_2/I_1} \tag{1.8}$$

where I_1 and I_2 represent the intensities of reflected beams from the two fiber end faces. The parameter V is often used to describe the contrast ratio of the interference fringe. When $I_2/I_1 = 1$, $V = 1$, and when $I_2/I_1 = 0$ or ∞, $V = 0$. When the interference of the two beams is the same ($I_1 = I_2 = I_0$), Equation 1.7 can be written as

$$I = 2I_0(1 + \cos\phi) = 4I_0\cos^2\left(\frac{\phi}{2}\right) \tag{1.9}$$

The typical optical spectrum of the two-beam interference is similar to a sinusoidal function, as shown in Figure 1.3b, according to Equation 1.9. When $\Delta = m\lambda$, where $m = 0, 1, 2, \ldots$, the interference intensity becomes maximum and if $\Delta = (m + 1/2)\lambda$, where $m = 0, 1, 2, \ldots$, the interference intensity becomes minimum. For fiber end faces, the reflectance is nearly 4%, so I_1 and I_2 are about 4% I_i and 3.69% I_i, and Equation 1.7 can be simplified as

$$I = 7.84\%\left(1 + \cos\frac{2\pi\Delta}{\lambda}\right)I_i \tag{1.10}$$

As Δ relies on the characteristics of the cavity, the changes of the cavity, such as length and refractive index, the optical spectrum will be shifted and can be used for sensing.

These kinds of FFPI sensors have simple structures and low cost. However, as the reflectivity of the FFPI cavity formed by two fiber ends is low, the optical power of reflected interferometric signal is normally small. In order to overcome the drawback of the two-beam FFPI, optical amplification (OA) can be effectively used to enhance the signal level [12]. The operating principle of such an FFPI/OA system is illustrated in Figure 1.4. Light from a broadband amplified spontaneous emission (ASE) source, consisting of an Er-doped fiber (EDF), a 1550/980-nm wavelength-division multiplexer (WDM) coupler, an isolator, and a pump laser, is launched into the FFPI. As shown in the enlarged view of the FFPI, a lead-in fiber and a reflecting

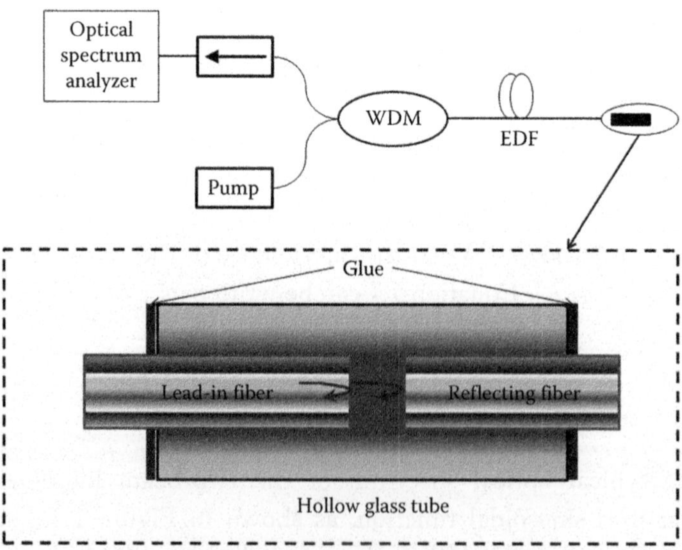

Figure 1.4 Two-beam FFPI sensing system with OA.

fiber are inserted into a hollow quartz tube. A certain air gap of several hundred micrometers between the two flat fiber ends is used to form the FP cavity. The interferometric signal from the FFPI is amplified by the EDF and then detected by an optical spectrum analyzer (OSA).

1.2.2 Three-Beam Interference

Generally speaking, the principle of three-beam interference is based on the interference of three reflective surfaces. Figure 1.5 shows one of the typical structures of the three-beam interference based on graded-index multimode fiber (GI-MMF) [13]. It is fabricated by cascading an air gap and a short section of GI-MMF to a singlemode fiber (SMF). The air gap can be formed by fusion splicing the SMF with a chemically etched micronotch on the GI-MMF end. The light propagation follows the sinusoidal path as the index profile is parabolic.

The ray-transfer-matrix (RTM) theory is used to describe the principle. A Gaussian beam can be expressed as

$$E(r,z) = \frac{A_0}{\omega} \exp\left(-i\Phi - \frac{inkr^2}{2q}\right) \tag{1.11}$$

where A_0 remains constant for energy conservation. $k = 2\pi/\lambda$ is the wave number λ with the free-space wavelength. $\Phi = nkz$ denotes the phase shift of the light beam as it propagates and n is the refractive

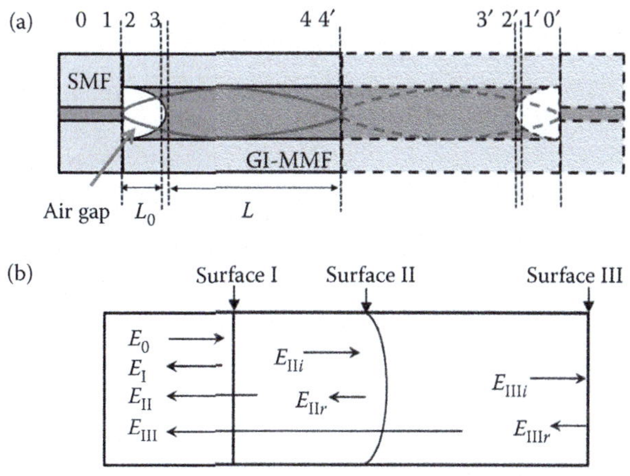

Figure 1.5 Three-beam FFPI cavity: (a) schematic and (b) principle.

index of the medium light propagating. The complex beam parameter, q, is given by

$$\frac{1}{q}=\frac{1}{\rho}-i\frac{\lambda}{n\pi\omega^2} \tag{1.12}$$

where ρ and ω are the radius of the curvature and the beam radius of the Gaussian beam, respectively. Suppose M is the transfer matrix from input plane to output plane, the transformation of the complex beam parameter is given by

$$q'=\frac{Aq+B}{Cq+D} \tag{1.13}$$

where $A=M(1,1)$, $B=M(1,2)$, $C=M(2,1)$, and $D=M(2,2)$, and q and q' are the complex beam parameters at the input and output planes, respectively.

The electrical amplitude of the incident beam at location 0 is $E_0(r)$ and corresponding complex parameter is $q_0=in_s p\omega_s^2/l$. n_s is the reflective index of the SMF core and ω_s is the beam radius of the SMF. The reflectance of surface R_{I} is determined by the Fresnel equation, $R_{\text{I}}=(n_s-n_0)^2/(n_s+n_0)^2$, and n_0 is the reflective index of the ambient medium, which is approximately 1 in the air. The electrical amplitude of the reflective beam at surface I is $E_{\text{I}}(r)=\sqrt{R_{\text{I}}}E_0(r)$. The reflectance of the etched micronotch (surface II) is represented by

$$R_{\text{II}}=\frac{\int_0^{\omega_{\text{II}r}}\left|E_{\text{II}r}(r)\right|^2 R(r)rdr}{\int_0^{\omega_{\text{II}i}}\left|E_{\text{II}i}(r)\right|^2 rdr} \tag{1.14}$$

where $\omega_{\text{II}i}$ and $\omega_{\text{II}r}$ are the beam radii of the incident light and the reflected light by surface II, respectively. $R(r)$ is the reflectance given by the radial distribution, $R(r)=[n(r)-n_0]^2/[n(r)+n_0]^2$, where $n(r)$ is the refractive index profile of the GI-MMF core, which can be given by

$$n(r)=n_1\sqrt{1-g^2r^2},\quad r<a \tag{1.15}$$

where n_1 is the maximum index at $r = 0$, a is the radius of the GI-MMF core, and g is a factor that determines the index profile of the core. The ABCD matrices corresponding to E_{IIr} and E_{IIi} are given by $M_{IIi} = M_{12}M_{01}$ and $M_{IIr} = M_2M_{12}M_{01}$, where M_{ij} is the matrix describing the transformation of the complex beam parameters between locations i and j. The elementary matrices are given by

$$M_{01} = \begin{bmatrix} 1 & 0 \\ 0 & n_s/n_0 \end{bmatrix}$$

$$M_{12} = \begin{bmatrix} 1 & L_0 \\ 0 & 1 \end{bmatrix}$$

$$M_2 = \begin{bmatrix} 1 & 0 \\ -2/\rho_1 & 1 \end{bmatrix}$$

where ρ_1 is the radius of curvature of the etched micronotch on the GI-MMF end and L_0 is the effective cavity length of the air gap, which is smaller than the distance between the SMF end and the bottom of the etched micronotch. The transformation of the complex beam parameters is determined by Equation 1.13. The electrical amplitude of the light beam reflected by surface II is $E_{II}(r) = \sqrt{R'_{II}}E'_{II}(r)$, where $R'_{II} = T_{II}^2\ (1 - A_I)\ R_{II}$. T_I is the transmittance of surface I and A_I is the propagation losses in the air gap. Φ_{II}, the additional phase in E_{II}, is given by $\Phi_{II} = 2n_0kL_0$. The ABCD matrix corresponding to E_{II} is presented as $M_{II} = M_{10}M_{21'}M_2M_{12}M_{01}$ with $M_{21'} = M_{12}$ and $M_{10} = \begin{bmatrix} 1 & 0 \\ 0 & n_0/n_s \end{bmatrix}$. The effective reflectance of surface II can be expressed as

$$R_{II_{eff}} = \frac{\int_0^{a_s} \left|E_{II}(r)\right|^2 rdr}{\int_0^{a_s} \left|E_0(r)\right|^2 rdr} \tag{1.16}$$

where $2a_s$ is the mode field diameter of the SMF.

Analogizing to the reflection R_{II}, the reflectance of surface III is given by

$$R_{III} = \frac{\int_0^{\omega_{IIIr}} |E_{IIIr}(r)|^2 R(r) r dr}{\int_0^{\omega_{IIIi}} |E_{IIIi}(r)|^2 r dr} \tag{1.17}$$

The electrical amplitude of the light beam reflected by surface III is $E_{III}(r) = \sqrt{R'_{III}} E'_{III}(r)$ with $R'_{III} = T_I^2 T_{II}^2 (1 - A_{II}) R_{III}$. T_I is the transmittance of surface I and A_I is the propagation losses in the air gap. Φ_{III}, the additional phase in E_{III}, is given by $\Phi_{III} = 2(n_0 L_0 + n_1 L)$. The ABCD matrices corresponding to E_{IIIi}, E_{IIIr}, and E_{III} are presented as $M_{IIIi} = M_{34} M_{23} M_{12} M_{01}$, $M_{IIIr} = M_{44'} M_{34} M_{23} M_{12} M_{01}$, and $M_{III} = M_{1'0'} M_{2'1'} M_{3'2'} M_{33'} M_{23} M_{12} M_{01}$, respectively, with

$$M_{23} = \begin{bmatrix} 1 & 0 \\ -(n_0 - n_1)/\rho_1 n_1 & n_0/n_1 \end{bmatrix}$$

$$M_{34} = \begin{bmatrix} \cos(gL) & \sin(gL)/g \\ -g\sin(gL) & \cos(gL) \end{bmatrix}$$

$$M_{33'} = \begin{bmatrix} \cos(2gL) & \sin(2gL)/g \\ -g\sin(2gL) & \cos(2gL) \end{bmatrix}$$

$$M_{3'2'} = \begin{bmatrix} 1 & 0 \\ -(n_1 - n_0)/\rho_1 n_0 & n_1/n_0 \end{bmatrix}$$

Considering the coupling losses of the light beam reflected from III into the SMF, the effective reflectance of surface III can be expressed as

$$R_{IIIeff} = \frac{\int_0^{a_s} |E_{III}(r)|^2 r dr}{\int_0^{a_s} |E_0(r)|^2 r dr} \tag{1.18}$$

R_{IIIeff} is mainly dependent on the coupling coefficient of the light beam into the SMF as the GI-MMF length changes. In general, R_{IIIeff} is smaller than R_{III}, because of the coupling losses and propagation losses. The reflective signal of the three-beam interference is given by

$$I(\lambda) = \int_0^{a_s} \left| E_I + E_{II} + E_{III} \right|^2 r dr \tag{1.19}$$

By using the effective reflectance of the three-beam interferometer, the above RTM theory can be simplified. For the case of $n_0 < n_1$, the normalized intensity of the three-beam interference can be expressed as

$$I(\lambda) = [R_I + R_{IIeff} + R_{IIIeff} + 2\sqrt{R_I R_{IIIeff}} \cos(\Phi_{III}) - 2\sqrt{R_I R_{IIeff}} \cos\Phi_{II} - 2\sqrt{R_{IIeff} R_{IIIeff}} \cos(\Phi_{III} - \Phi_{II})] I_0 \tag{1.20}$$

It is well known that the fringe contrast of the two-beam interference becomes maximum when the reflectance of the surfaces is equal, which is a strict constraint. The corresponding constraint condition on the effective reflectance of the three surfaces for the three-beam interference can be deduced by Equation 1.10 and is given by

$$\left| \frac{(R_I + R_{IIeff} - R_{IIIeff})}{(2\sqrt{R_I R_{IIeff}})} \right| \le 1 \tag{1.21}$$

Unlike the two-beam interference, the constraint condition for the three-beam interference to obtain the optimal fringe contrast is an inequality; that is, the requirement on the reflectance is a relatively wide range rather than a decided value. This makes it easier for sensing based on three-beam interference to obtain high performance than the conventional two-beam interference. Figure 1.6 shows the reflective spectra of the SMF end (black), the air gap (light gray), and the three-beam interferometer introduced above with GI-MMF length of 515 μm (gray) [14].

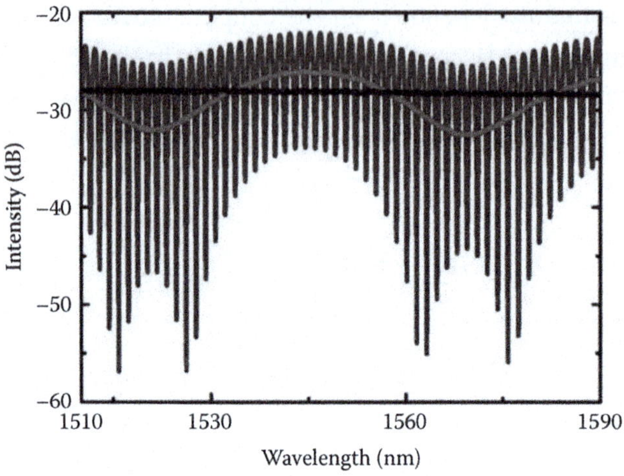

Figure 1.6 Reflective spectra of two- and three-beam interferences.

1.2.3 Multi-Beam Interference

The interferences above are actually approximate results based on the reflecting planes with low reflectance. For example, the reflectance of the fiber end is about 4%. It is suitable as concerns about only two beams according to the reason introduced in Section 1.2.1. If the reflectance of the reflecting plane increased, the result will be different. Similarly, if the intensity of the incident light is 100%, but the reflectance is 90%, the intensity of the first transmitted light is 1%, the second is 0.81%, the third is 0.66%, and so on. Although the intensities of these transmitted lights are decreased, the difference between the adjacent light is little enough. As a consequence, multi-beam interference has to be considered in this situation.

Consider the light shown successively reflected between the fiber ends of the FP cavity A and B in Figure 1.7. The incident ray *i* of unit amplitude stands for the propagating direction of the incident light and θ represents the incident angle in the cavity as shown. λ_0 is the wavelength in vacuum, and the effective refractive index of the medium in the FP cavity is n. For convenience, the electric vector of the incident light is considered as unit amplitude, and linearly polarized either parallel or perpendicular to the incident plane; the OPD between two light beams is

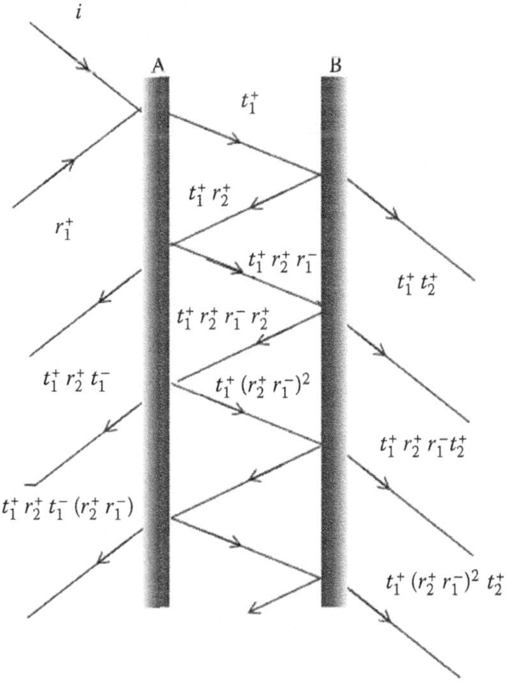

Figure 1.7 Multi-beam interference.

$$\Delta = 2nh\cos\theta \tag{1.22}$$

The corresponding phase difference is

$$\varphi = \frac{2\pi\Delta}{\lambda} = \frac{4\pi}{\lambda} nh\cos\theta \tag{1.23}$$

It is supposed that E_{0i} is the amplitude of incident light electric vector, and r, r', t, and t' are the reflection factors and the transmission factors when light goes in and out of the cavity. So, the amplitudes of the light reflecting from the FP cavity are

$$E_{01r} = rE_{0i}$$

$$E_{02r} = r'tt'E_{0i}\exp(i\phi)$$

$$E_{0lr} = tt'r'^{(2l-3)}E_{0i}\exp[i(l-1)\phi]$$

The complex optical vector of P is

$$\begin{aligned} E_{0r} &= E_{01r} + \sum_{l=2}^{\infty} E_{0lr} \\ &= E_{01r} + \sum_{l=2}^{\infty} tt'r'^{(2l-3)} E_{0i} \exp[i(l-1)\phi] \\ &= E_{01r} + tt'r'E_{0i} \exp(i\phi) \sum_{l=0}^{\infty} r'^{2l} \exp(il\phi) \end{aligned} \tag{1.24}$$

According to the Fresnel formula, r, r', t, and t' satisfy the relations:

$$r = -r'$$

$$tt' = 1 - rr' = 1 - r^2$$

The reflectance R and the transmittance T also have relations:

$$r = -r'^2 = R$$

$$tt' = 1 - R = T$$

Using the mathematical equation,

$$\sum_{n=0}^{\infty} x^n = \frac{1}{1-x}$$

If the quantity of the reflected beams is big enough, Equation 1.24 can be simplified as

$$E_{0r} = \frac{[1 - \exp(i\phi)]\sqrt{R}}{1 - R\exp(i\phi)} E_{0i} \tag{1.25}$$

According to the equation above, the intensity of the reflected light after focusing is

$$I_r = \langle E^2 \rangle = \frac{F \sin^2(\phi/2)}{1 + F\sin^2(\phi/2)} I_i \tag{1.26}$$

$$F = \frac{4R}{(1-R)^2} \tag{1.27}$$

Usually, the coefficient F can reflect the finesse of the interference fringe, which will be explained as detailed below.

Similarly, the intensity of the transmitted light after focusing is

$$I_r = \frac{T^2}{(1-R)^2 + 4R\sin^2(\phi/2)} I_i \quad (1.28)$$

Ignoring the absorption, $T = 1 - R$, the transmitted light can be simplified as

$$I_t = \frac{1}{1 + F\sin^2(\phi/2)} I_i \quad (1.29)$$

Equations 1.26 and 1.29, the intensity distribution formulas of the interferences by the reflected light beams or the transmitted light beams, are usually named *Airy formulas*. We also know $I_r + I_t = I_i$, which reflected the universal rule of energy conservation. If the reflected light is increased by interference, the transmitted light will be decreased and vice versa. It means that the distributions of the reflected light and the transmitted light are complementary.

According to Airy formulas, for reflected light, if $\phi = (2m + 1)\pi$, $m = 0, 1, 2, \ldots$, the maximum intensity, I_{rM} satisfies $I_{rM} = (F/1 + F)I_i$ and if $\phi = 2m\pi$, $m = 0, 1, 2, \ldots$, the minimum intensity, I_{rm} satisfies $I_{rm} = 0$. Similarly, the maximum and minimum of the transmitted light occur when $\phi = 2m\pi$ and $\phi = 2m\pi$, $m = 0, 1, 2, \ldots$, separately, with $I_{tM} = I_i$ and $I_{tm} = (1/1 + F)I_i$. Comparing with Section 1.2.1, that is, the two-beam interference, the conditions of the max or min interference intensities for reflected light or transmitted light are the same for both two-beam interference and multi-beam interference.

With proper spacing of the reflectors, the transmittance of the FP cavity is high. Changing the spacing causes transmittance to drop. With high-reflectivity reflectors, the transmittance is very sensitive to changes in wavelength or reflector spacing. Overall, interferometer performance is frequently characterized by finesse F related by the reflectance R, shown in Equation 1.27. Transmission as a function of reflector spacing is shown in Figure 1.8 for various values of finesse, in which the x-coordinate presents the phase difference of the adjacent transmitted light and the y-coordinate presents the relative intensity. High-finesse interferometers are useful because of the precisely located spectral features; low-finesse interferometers provide linear operation with a wider range of the measurand without complex feedback schemes.

Taking the formulas of I_{rM}, I_{rm}, I_{tM}, and I_{tm}, back to Equation 1.7, the contrast ratios of reflection and transmission are

$$V_r = 1 \tag{1.30}$$

$$V_t = \frac{F}{(2+F)} = \frac{2R}{(1+R^2)} \tag{1.31}$$

Obviously, V_t is less than 1; however, it does not mean that the transmitted spectrum is always worse than the reflected spectrum. As shown in Figure 1.8, the larger the R is, the sharper the transmitted spectrum will be. According to the relationship of I_r, I_t, and I_i, $I_r + I_t = I_i$, the reflected spectrum will be wider if R is larger.

We can find that the distribution of the light intensity is corresponding to the reflectance R. If R is really small, F is much less than 1. If carrying out Equations 1.25 and 1.27 and only reserving

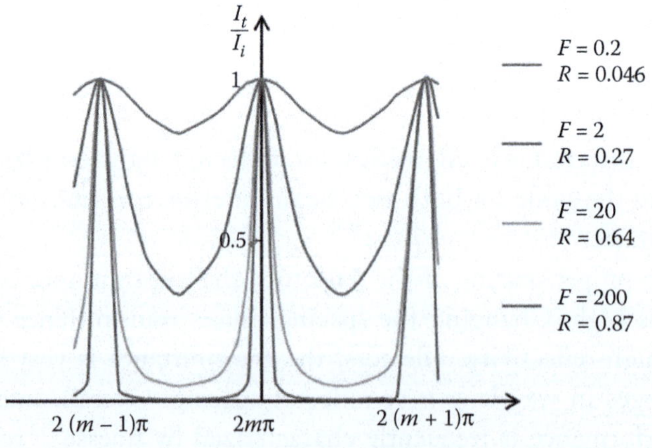

Figure 1.8 Transmission spectrum of multi-beam interference.

the first item, the relative intensities of reflected light and transmitted light are

$$\frac{I_r}{I_i} \approx F\sin^2\frac{\phi}{2} = \frac{F}{2}(1-\cos\phi) \tag{1.32}$$

$$\frac{I_t}{I_i} \approx 1 - F\sin^2\frac{\phi}{2} = 1 - \frac{F}{2}(1-\cos\phi). \tag{1.33}$$

It can be proved that the equations above are just the intensity distributions of the two-beam interference.

The finesse of the spectrum is also relative to R. Higher finesse of the interference spectrum by transmitted light is just the most important factor of the multi-beam interference.

The finesse of the transmitted interference light can be reflected by the half-high width $\Delta\varphi$ introduced in Section 1.2. As shown in Figure 1.9, the phase of the half-high position is $\phi = 2m\pi \pm (\Delta\phi/2)$. Taking it back to Equation 1.27, we can get an equation

$$\frac{1}{1+F\sin^2(\Delta\varphi/4)} = \frac{1}{2} \tag{1.34}$$

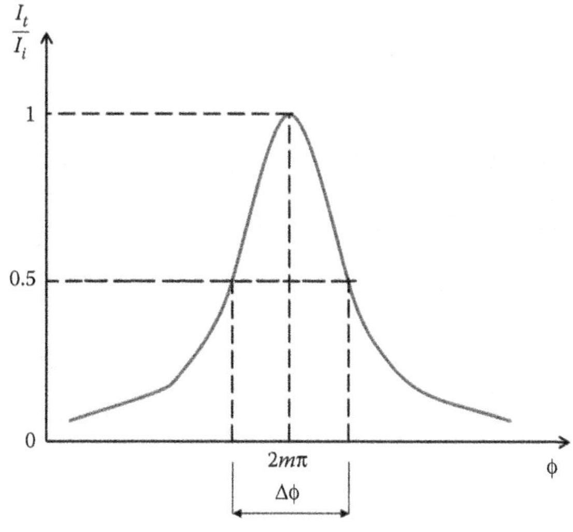

Figure 1.9 Half-high width $\Delta\varphi$.

If $\Delta\varphi$ is really small, $\sin(\Delta\varphi/4) \approx \Delta\varphi/4$. The equation of the half-high width $\Delta\varphi$ can be derived, which is already shown in Equation 1.5.

References

1. Born, M. and Wolf, E. 1980. *Principles of Optics* (6th ed.). Oxford: Pergamon.
2. Hernández, G. 1988. *Fabry–Perot Interferometers* (Vol. 3). Cambridge: Cambridge University Press.
3. Fabry, C. and Perot, A. 1899. Theorie et applications d'une nouvelle methode de spectroscopie interferentielle. *Annales de Chimie et de Physique*, 16(7), 115.
4. Franzen, D. L. and Kim, E. M. 1981. Long optical-fiber Fabry–Perot interferometers. *Applied Optics*, 20(23), 3991–3992.
5. Cielo, P. G. 1979. Fiber optic hydrophone: Improved strain configuration and environmental noise protection. *Applied Optics*, 18(17), 2933–2937.
6. Yoshino, T. and Ohno, Y. 1981. Fiber Fabry–Perot interferometers. In *Optical Fiber Communication Conference* (Optical Society of America, p. WL2), San Francisco, California, USA.
7. Petuchowski, S. J., Giallorenzi, T. G., and Sheem, S. K. 1981. A sensitive fiber-optic Fabry–Perot interferometer. *IEEE Journal of Quantum Electronics*, 17, 2168–2170.
8. Udd, E. 1996. Fiber optic smart structures. *Proceedings of the IEEE*, 84(6), 884–894.
9. Lee, C. E. and Taylor, H. F. 1995. *Sensors for Smart Structures Based on the Fabry–Perot Interferometer. Fiber Optic Smart Structures (A 95-34976 09-39), Wiley Series in Pure and Applied Optics* (pp. 249–269). New York, NY: John Wiley & Sons, Inc.
10. Perot, A., Fabry, C., Perot, A., and Fabry, C. 1898. Sur un voltmètre électrostatique interférentiel pour étalonnage [An electrostatic voltmeter by calibration of interference]. *Journal de Physique Théorique et Appliquée*, 7(1), 650–659.
11. Meggers, W. F. and Peters, C. G. 1918. Measurements on the index of refraction of air for wave-lengths from 2218a to 9000a. *Journal of the Franklin Institute*, 186(1), 747–749.
12. Rao, Y. J. 2006. Recent progress in fiber-optic extrinsic Fabry–Perot interferometric sensors. *Optical Fiber Technology*, 12(3), 227–237.
13. Gong, Y., Zhao, T., Rao, Y. J., Wu, Y., and Guo, Y. 2010. A ray-transfer-matrix model for hybrid fiber Fabry–Perot sensor based on graded-index multimode fiber. *Optics Express*, 18(15), 15844–15852.
14. Gong, Y., Guo, Y., Rao, Y. J., and Zhao, T. 2010. Fiber-optic Fabry–Pérot sensor based on periodic focusing effect of graded-index multimode fibers. *IEEE Photonics Technology Letters*, 22(23), 1708–1710.

2 Microstructures of FFPI Sensors

Free-space FP cavities have been universally employed for applications in lasers, spectroscopy, and filters, just to name a few. There are however different types of FP structures, where the stable optical resonator consisting of two concave mirrors is most often used. Both free-space mirrors and the cavity itself are bulky, centimeters in diameter and tens of centimeters in length, and it is necessary to use the stable resonator to minimize the propagation loss and to achieve a high Q factor, which contributes much to the continuous development of high-power lasers, as well as getting longer light–matter interaction length in cavity ring-down spectroscopy.

Compared with conventional free-space FP devices, the development of fiber FP interferometers (FFPIs) represents a big step, not only in terms of great miniaturization but also in terms of the enrichment of the various structures, thanks to the variety of optical fibers. Actually, it already forms one of the most important directions in the field of optical fiber sensors, which is reflected in the large number of publications and the commercial product for practical use. In this chapter, we introduce various microstructures of FFPIs and their corresponding advantages.

There are several ways to classify FFPI sensors. First, in terms of fiber types, it includes FFPIs based on singlemode and multimode, as well as microstructured fibers, tapered or microfibers, etc. Second, they can be classified into intrinsic, extrinsic, and hybrid FFPIs. Usually, the two reflective surfaces in intrinsic FFPIs are separated by a section of singlemode fiber (SMF), while it is an air gap which forms the cavity of the extrinsic FFPIs (EFFPIs). Third, FFPI sensors can be classified in terms of Q factor or finesse.

2.1 Singlemode FFPI Structures

Among all kinds of optical fibers, the SMF has the lowest transmission loss and also the lowest cost, thanks to the development of optical fiber communications, which make it an excellent candidate for the fabrication of FFPIs. One can use the fiber Bragg grating (FBG) pairs, reflective films to provide multi-beam interference, or the Fresnel reflection between the silica material of optical fibers and air to perform the two-beam interference, for the FFPIs. The principle of these FFPI sensors was described in Chapter 1.

2.1.1 Intrinsic FFPI Sensors

The intrinsic and extrinsic FFPIs (IFFPI, EFFPI) are differentiated by whether the material is the same as the fiber (silica) between the reflective surfaces. The IFFPIs use the same material, whereas the EFFPIs use different materials. The IFFPI sensors often consist of two reflective elements and a section of SMF in between. The reflective elements can be FBG pairs, reflective films, air holes, or surfaces formed by fusion splicing between different kinds of optical fibers.

2.1.1.1 IFFPI Structures Based on FBG Pairs

2.1.1.1.1 Fundamentals of FBG FBG was first demonstrated by Kawasaki et al. [1] in 1978 and has been widely used for optical fiber communications, fiber sensors, and fiber lasers. Typical applications of the FBGs include filters, temperature and strain sensors, and laser cavity mirrors. FBG sensors have been developed to be one of the largest markets of all types of optical fiber sensors, especially for structure health monitoring [2,3].

By using an ultraviolet (UV) laser, FBGs can be inscribed in the fiber core by introducing periodic refractive index changes. Usually, photosensitive fibers or hydrogen-loaded fibers are used in order to improve the performance of FBGs. There are several techniques often used for the fabrication of FBGs. One is by using a pulsed excimer laser, at a wavelength of 248 nm, and a phase mask. The coherence length of the excimer laser is often at the 1 mm level or even below. The interference occurs right behind the phase mask so that there is no requirement of a long coherence length on the laser and the optical fiber is put close to the phase mask [4]. This method is most often used for commercial production of FBGs due to its high efficiency for fabrication and high stability of the fabricated FBGs.

The other is by using the two-beam interference method based on the 244 nm continuous wave (CW) Ar^+ laser [5], whose coherent length is much longer than the excimer laser. This method is flexible and can be easily designed for generating FBGs with a desired wavelength by finely tuning the interference between the two beams.

There is also the point-by-point inscription method for the fabrication of FBGs [6,7]. It needs motorized moving parts to scan the laser spot and can be used for the fabrication of chirped or apodized FBGs with a specially designed refractive index profile.

The basic principles of FBGs can be found in the literature. The Bragg wavelength can be expressed as $\lambda = 2n\Lambda$, where Λ is the grating pitch and n is the effective index of the core [8]. The grating pitch is determined by the half period of the phase mask or the period of the two-beam interference. FBGs are a type of reflective element and often used as a narrow bandpass filter, thanks to their low insertion loss. Usually, the FBG with uniform index changes has a reflective bandwidth of about 0.2 ~ 0.4 nm and the bandwidth can be extended into several or more than 10 nm for a chirped FBG, whose period of the index changes is not uniform [9].

2.1.1.1.2 Principle of IFFPIs Based on FBGs

2.1.1.1.2.1 Low-Reflectance IFFPIs Based on FBGs By using a pair of FBGs with a reflective band, IFFPI sensors can be formed. Wan and Taylor proposed an IFFPI sensor based on FBG pairs for temperature measurement [10]. Two short sections, 2 mm in length, were cut from a 2-cm-long FBG and spliced with a SMF in between. The cavity length is 15 mm. The original reflectance of the 2-cm-long FBG is almost 100% and that of the 2-mm FBG mirrors is ~2%. The reflected beams from the FBG pairs interference and fringes can be detected.

Due to the low reflectance of the FBG pairs, the reflection of the IFFPI sensor can be considered by the two-beam interference model, similar as shown light gray curve in Figure 1.6. The reflectance of such an IFFPI sensor can be expressed as

$$R = r_1 + r_2 + 2(r_1 r_2)^{1/2} \cos\phi \tag{2.1}$$

where r_1 and r_2 are the reflectance of the two FBGs. $\phi = 4\pi nL/\lambda$ is the round-trip phase shift, where n is the effective refractive index of the fiber mode, L is the cavity length of the IFFPI, and λ is the wavelength. The use of weak FBGs for forming the IFFPI sensor is beneficial for multiplexing.

By using a wavelength-swept distributed feedback (DFB) laser, the wavelength and the peak intensity of the interference fringes can be determined as a function of the ambient temperature. The temperature was measured in the range of 25–170°C with a resolution of 0.005°C.

2.1.1.1.2.2 High-Reflectance IFFPIs Based on FBG The performance of the IFFPI sensor based on FBG may degrade if there is a mismatch in the spectral profile of the weak FBG pairs. Niu et al. [11] investigated the performance of the IFFPI sensor by using high-reflectance FBGs, that is, strong FBGs. The incident beam was reflected by multiple times between the two strong FBGs, which can amplify the effect of fiber strain on the phase changes of the beam. Therefore, high sensitivity can be achieved by using the high-reflectance IFFPI sensors. Furthermore, the linewidth of the interference fringes becomes narrower and is beneficial for high-resolution spectral measurement. The transmission spectra of the high-finesse IFFPI sensors can be found in Figure 1.8.

2.1.1.1.2.3 Advantages of FBG-IFFPIs: High Mechanical Strength, Multiplexing, Absolute Measurement, and Large Dynamic Range IFFPI sensors based on FBG can stand strain of up to 12,000 $\mu\varepsilon$ as the FBG inscription process does not require any cleaving or splicing over the sensor region [12]. IFFPI sensors based on low-reflectance FBG pairs can be used for multiplexing. A total of 50 sensors were experimentally multiplexed by the frequency-division multiplexing method, and the capacity was theoretically estimated to be 500 sensors [13]. Compared with other types of IFFPI sensors, IFFPI sensors based on FBG have the advantages of low insertion loss and small spectral width for each sensor, which greatly enhance the multiplexing capability of IFFPI sensors.

These sensors combine the advantages of both fiber-optic interferometric sensors and FBG sensors. Fiber-optic interferometric sensors often provide high resolution based on the phase-shift detection. However, they suffer from the ambiguity problem that the absolute measurement is not achievable by detecting only one interferometric wavelength, when using the broadband reflections such as the Fresnel reflection or other highly reflective coatings. If the wavelength shift is constrained within a single free spectral range (FSR), the dynamic range is small. On the other hand, FBG sensors have no ambiguity problems by detecting the unique Bragg wavelength, but the measurement resolution is about 1 $\mu\varepsilon$, which is limited by the wavelength interrogation method and the relatively wide 3 dB linewidth. The ratio of the dynamic range to resolution is around 10^3:1.

Rao et al. demonstrated an IFFPI sensor based on FBG for absolute strain measurement [14]. The bandwidth of the FBGs for forming the IFFPI cavity was ~0.5 nm and the reflectivity was 10%. The wavelength shift of the FBG versus strain was measured with an optical spectrum analyzer with a resolution of 1 pm. The phase change of the IFFPI cavity was detected by a lock-in amplifier. The Bragg wavelength shift is used for determining the fringe number, N, so that the absolute phase change of the IFFPI, $\Delta\Phi + 2\pi N$, is determined for the absolute strain sensing. By combining the advantages of both FBG and interferometric sensors, the ratio of the dynamic range to resolution is extended up to 2.4×10^4:1.

Compared with IFFPI with a single pair of Bragg gratings, the identification of the center of the fringe pattern is easier by multiple co-located pairs of FBGs and low-coherence interrogation [15]. FBG-IFFPI sensors can also be designed for dual-parameter measurement [16].

Chirped FBGs (CFBGs) have a wide reflection band of several to tens of nanometers [17], due to the nonuniform refractive index profile in the fiber core. IFFPI sensors can be formed by using two identical CFBGs. There are many transmittance peaks within the reflection band of the sensor. By using a pair of CFBGs with a linewidth of 20 nm and a narrow wavelength sweep range of 0.2 nm, the dynamic range of the sensor can be extended by 100 times compared with that of IFFPI sensors based on FBG. The linewidth and FSR of the IFFPI sensor based on CFBG were 5 and 65 pm, respectively [18].

2.1.1.2 IFFPI Structures Based on Reflective Films Highly reflective mirrors have been widely used for constructing free-space, high-finesse

FP cavities. The same strategy was employed for IFFPI sensors. One of the most promising advantages of coatings on optical fiber is mass production. Thousands of optical fibers can be coated at one time with the same reflective spectral performance. The spectral characteristics of the films can be designed by the mature thin-film optics method. However, it is still challenging to fusion splice the coated fiber with another cleaved fiber to form an in-line reflective mirror. The reflective spectra of the film may distort and the reflectance may decrease greatly during the fusion splicing.

On the other hand, the cross-sectional area of optical fibers is small, with a diameter of ~125 μm. Dielectric multi-layer stack of optical coatings with a thickness of several micrometers is not as stable as that coated on a free-space substrate, when coated on a cleaved surface of an optical fiber. Therefore, the thinner, the better the reflective films coated on the optical fiber , as long as the reflective characteristics are preserved.

In 1988, H. Taylor and coworkers proposed an IFFPI temperature sensor by introducing a pair of dielectric mirrors into the singlemode optical fiber [19]. Cleaved fibers were coated with TiO_2 film with a thickness of 100 nm by the dc magnetron sputtering system. Then the coated fiber was fusion spliced to another cleaved uncoated fiber to form an in-line mirror. The reflectance of such a mirror was 2% and the cavity length is 1.5 mm. The insertion loss of each mirror was reduced to 0.1–0.2 dB by positioning the fiber during the splicing. The sensor was used for temperature sensing from –200°C to 1050°C as the materials, both SiO_2 and TiO_2, in this structure can withstand high temperature [20]. The reflective film can be fabricated by the vacuum deposition, magnetron sputtering, or e-beam evaporation [21].

Recently, chromium (Cr), with a melting point of 1907°C was used, instead of TiO_2 (1843°C), as the reflective mirror for fabricating the FFPI sensor [22], as shown in Figure 2.1. This melting temperature is slightly higher so that the long-term stability of the sensor is improved. The interference between the coated Cr film and the Fresnel reflection was used as the read-out signal of the IFFPI sensor. A Cr film with a reflectance of ~10% was coated on the cleaved fiber end. When spliced with another cleaved fiber, the reflectance decreases to ~4%, which is similar to that of the Fresnel reflection so that the fringe contrast is maximized. Temperature up to 1100°C was

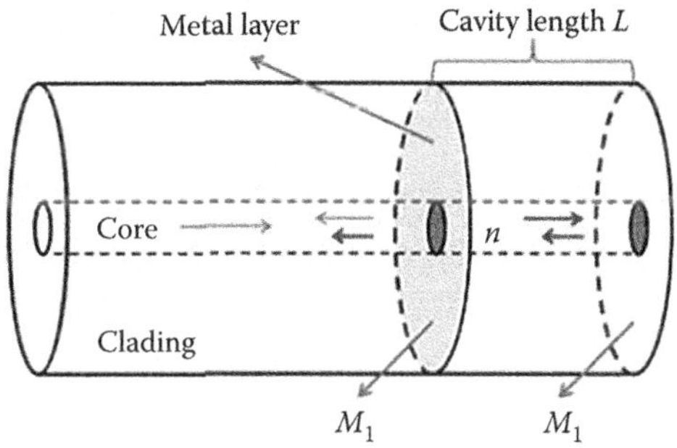

Figure 2.1 Schematic of the reflective film-based IFFPI sensor.

measured with a resolution of 4°C. The 300-hour long-term stability was estimated to be about 10°C.

2.1.1.3 IFFPI Structures Based on Air Holes Besides introducing in-line FBGs or reflective films as mirrors, IFFPI sensors can be constructed by using an air hole as the reflective element. The Fresnel reflection between the fiber material and air offers a reflection of about 4%, which is sufficient to form a high-performance mirror for the IFFPI sensor based on two-beam interference.

The fabrication process of the micro-air-hole-based IFFPI sensors includes three steps: drilling a micro-hole at the fiber end face, fusion splicing the fabricated fiber with another cleaved fiber to form the air gap, and cleaving, as shown in Figure 2.2. The last cleaving step makes another Fresnel reflection for the two-beam interference for the IFFPI sensor.

There are several methods to make a micro-hole at the fiber end face, including chemical etching, laser micromachining, focused ion beam (FIB) milling, etc. All these methods have been used for

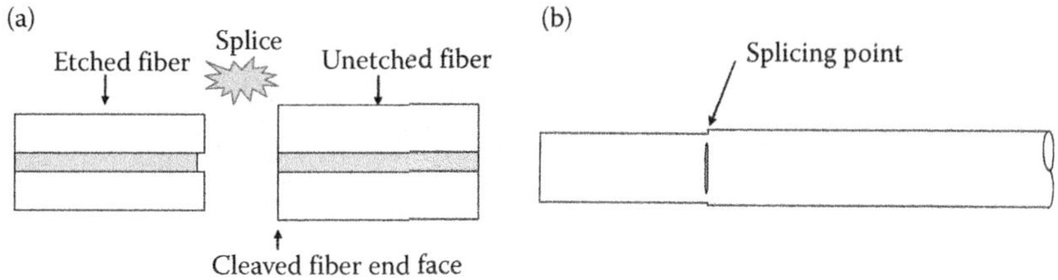

Figure 2.2 Fabrication process of the air-gap mirror for the IFFPI sensors. (a) Splicing etched fiber with a cleaved fiber, (b) the fabricated air-gap mirror.

fabricating a wide air gap to form an extrinsic IFFPI sensor, which will be introduced in Section 2.1.2. For the intrinsic IFFPI sensor, the air gap serves as one reflective mirror so that the width should be much narrower than that used in the EFFPI sensor. The most convenient and cost-effective way is by using wet chemical etching. It is also good for mass production. It was used in the 1990s for fabricating sharp fiber tapers at nanoscale for near-field scanning optical microscopy [23,24].

By using the hydrofluoric (HF) acid, the etching rate is faster for the Ge-doped silica fiber core, compared with that of the F-doped fiber cladding. Therefore, a micro-air-hole can be fabricated at the fiber end due to the etching rate difference. The etching rate depends on the concentration of HF acid. So a cover layer of nonvolatile reagent that is insoluble in water can be used to prevent the concentration changes due to volatilization of HF. Buffered HF acid, that is, a mixture of HF acid and ammonium fluoride (NH_4F), was often used to control the etching rate. However, NH_4F is not necessary, and only changing the concentration of HF acid is more than enough to control the etching rate. Note that higher concentration of HF acid leads to a rougher surface, along with a higher etching rate. Therefore, a better choice is to use low concentration of HF acid and long etching time to obtain a smooth-etched surface.

Tso and Pask investigated the influence of the concentration of HF acid on the etching rate and found that the reaction rate increases in a linear fashion when the concentration is lower than 10 M or higher than 25 M [25]. The diffusion process influences the etching rate and agitation during the etching process would increase the etching rate. The dissolution rate, and thus the etching rate, also increases with an increasing temperature [26].

The second step is to fusion splice the etched fiber to a cleaved fiber. An air gap is formed during the splicing and the whole air gap acts as one of the reflective mirrors for the IFFPI sensor. Anbo Wang and his coworkers investigated the influence of the shape, including the width and the diameter, of the air gap on the transmittance characteristics [27]. They suggested that the width of the air gap of $d < 3$ μm and a diameter of more than 20 μm are required for the highest coupling coefficient, as shown in Figure 2.3. They demonstrated two structures for the IFFPI sensors. The first is by using one air-gap mirror and one Fresnel reflection of a cleaved fiber. It is good for single-point

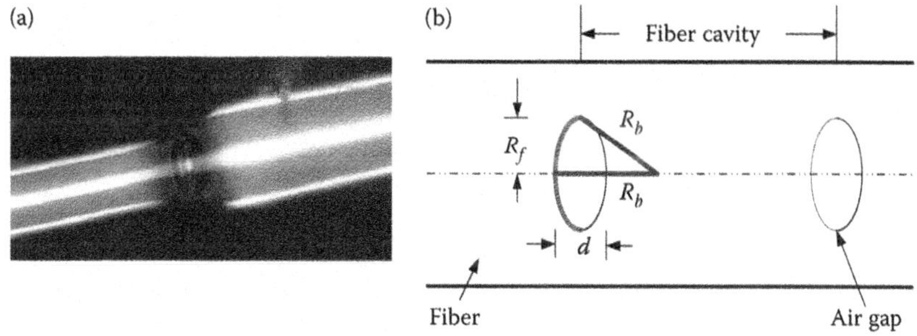

Figure 2.3 (a) Microscopic image of the air-gap mirror and (b) schematic of the IFFPI sensor based on air-gap mirrors.

measurement. The second is by using two air-gap mirrors. It is better for multi-points measurement via sensor multiplexing. Both temperature and strain measurement were carried out by these air-gap-based IFFPI sensors.

Fernando and his coworkers investigated the chemical etching of optical fibers in detail by using a 3D surface profiler, optical microscopy, and scanning electron microscopy [28]. Although the authors claimed that the sensor was an intrinsic FFPI sensor, it is classified as a EFFPI sensor in this book, as the medium in the cavity is air rather than a section of fiber. However, the method described in their work is very helpful to fabricate the air-gap mirror for intrinsic FFPI sensors. The concentration of 48% HF acid was used for the etching. No NH_4F was used. The etched fiber diameter, etch depth, and the cone angle of the etched hole were recorded versus etching time. A decrease of the fiber diameter in a linear fashion at about 3 μm/min was observed. The depth of the etched hole increased linearly with a rate of 1 μm/min. Their experiment was performed at a temperature of 20°C with a stable water bath. The roughness was around 21–44 nm with an etching time of 5–16 min, but increased up to 147 nm when etched for 22 min.

The FFPI sensor was then fabricated by fusion splicing two etched fibers. One advantage of splicing two etched fibers is that there is no difference between the diameter of the two etched fibers so that the automatic alignment during the splicing process is more stable, and moreover, the mechanical strength is better. In the case of splicing one etched fiber with a cleaved fiber, the fiber diameter is different and may cause weak mechanical properties. However, the fiber diameter is small with an etching time longer than 20 min and it may require precise alignment and careful choice of the fusion splicing parameters. Generally speaking, after chemical etching, the melting point of the fiber may decrease. Deformation and expansion may occur at the splicing part.

2.1.1.4 IFFPI Structures Based on Fusion Splicing of Different Kinds of Fibers The main point in the fabrication of IFFPI sensors is to form reflective mirrors in the optical fiber, in order to provide at least two reflective beams for generating the interferometric signal. Previously, FBG pairs, reflective films, and air gaps were introduced as reflective mirrors. For all three structures, it is necessary to fabricate the reflective elements first and then to form the FP cavity based on the two reflective mirrors. Indeed, there is another simpler method to fabricate IFFPI sensors, that is, by directly fusion splicing two different kinds of optical fibers. As long as there is a refractive index difference between the two fibers, there is a reflective signal according to the Fresnel equation. By considering the light with an incident angle of 90°, the Fresnel equation can be simplified and the reflection can be expressed as

$$R = \left(\frac{n_1 - n_2}{n_1 + n_2} \right)^2 \tag{2.2}$$

Here n_1 and n_2 are the refractive indices of the two optical fibers.

The first example is the work by Tsai and Lin published in 2001 [29]. They used Corning SMF (SMF-28) and the 3M SMF (FS-SN-3224), respectively. The core diameters were 8.4 and 4 μm, respectively. The refractive indices of the fiber cores were 1.4488 and 1.456 at 1550 nm. Due to the core diameter difference, the reflectance at the spliced joint of the two fibers depended on the core refractive index difference within the radial range of 0–2 μm, as well as the refractive index difference of the core of SMF-28 and the cladding of FS-SN-3224 within the radial range of 2–4.2 μm. They used a splicing point as one reflective mirror, and a cleaved fiber end as another reflective mirror, as shown in Figure 2.4. The fringe contrast of the interference may not be excellent due to the different reflectance from the two mirrors.

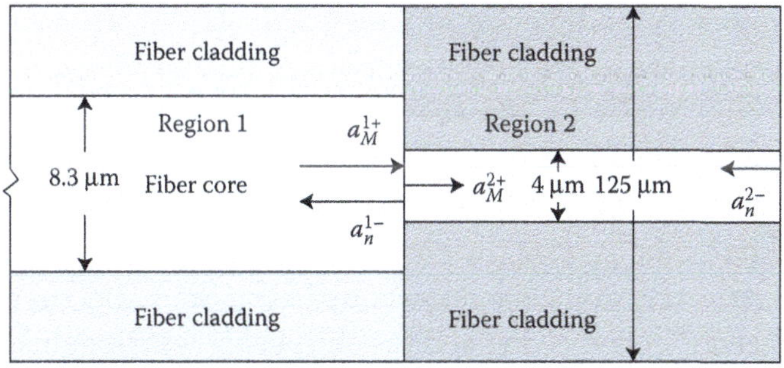

Figure 2.4 Structure of an FP cavity with different fiber cores.

Wang and coworkers [30] proposed another IFFPI sensor by using two identical mirrors based on splicing a section of a multimode optical fiber in two SMFs, that is, the singlemode–multimode–singlemode (SMS) structure, as shown in Figure 2.5. The fibers were Corning SMF-28 SMF and Corning InfiniCor 50-μm MMF. The cavity length was controlled with an accuracy of ±70 μm. The insertion loss was not a problem. When the light propagating in the SMF is launched into the MMF, it primarily excites the fundamental mode of the MMF.

By detecting the reflected light from the SMS structure, the interference fringes were recorded. The fringe contrast was about 15 dB, good enough for most physical parameter sensing. The average reflected light intensity was about –50 dB; compared with that, the Fresnel reflection was –13.8 dB. Therefore, the sensing signal is rather small. The authors demonstrated that the insertion loss was about 0.90 dB with a standard deviation of 0.17 dB by fabricating 50

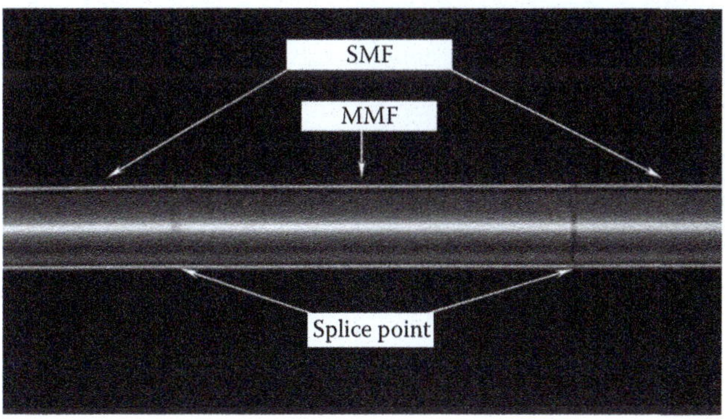

Figure 2.5 IFFPI sensor based on the SMS structure.

such sensors. A multiplexing of 8 such SMS-IFFPI sensors was also demonstrated with cavity lengths between 553 and 5332 μm.

Recently, Xuejin Li and coworkers further developed an IFFPI structure based on a section of microstructured optical fiber (MOF) fusion spliced to the SMF [31], as shown in Figure 2.6. The partial Fresnel reflection is from the interface of the two fibers and end face Fresnel reflection of the microstructured fiber. Similar to the previous two schemes, only cleaving and fusion splicing were required for the fabrication of such IFFPI structure. The fringe contrast can be as high as 20 dB when the microstructured fiber length is about 50 μm. Due to the trend of the fringe contrast versus the fiber length, the fringe contrast can be further enhanced if shorter fiber length is used. The high-temperature sensing performance is good. Temperature up to 1000°C was tested and a sensitivity of 17.7 pm/°C was obtained.

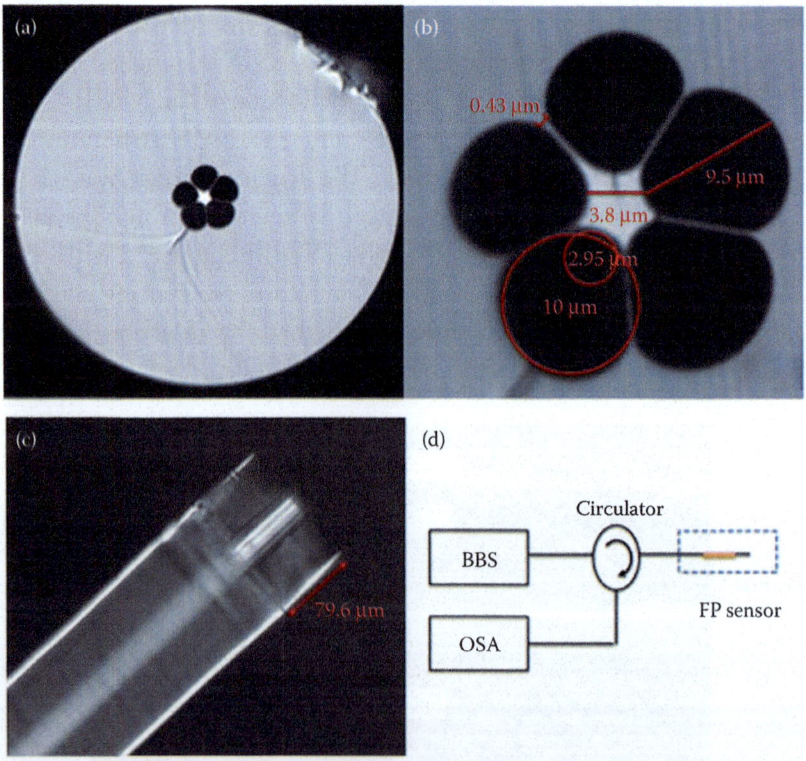

Figure 2.6 IFFPI structure based on splicing a section of microstructured fiber to SMF. (a) Cross-section of the microstructured fiber and (b) detailed dimensions. (c) The fabricated IFFPI structure and (d) its schematic interrogation setup.

The most promising point of this kind of IFFPI sensor is the simple fabrication process; that is, only cleaving and fusion splicing are required. And it is good for high-temperature sensing as the whole structure was based on silica.

2.1.1.5 Asymmetric IFFPI Structures We have already introduced several types of reflective mirrors for constructing the IFFPI structures by using a pair of such mirrors, including FBGs, reflective films, and Fresnel reflections. The Fresnel reflection mirrors include three different structures: (1) the air gaps by fusion splicing a chemically etched fiber and a cleaved fiber, (2) the reflection from the splicing joint of two different kinds of optical fibers, and (3) a cleaved fiber end face. Theoretically speaking, IFFPI structures can be formed by choosing either two of these reflective mirrors. Here, the asymmetric IFFPI structures can be classified into two types. One is asymmetric in the axial direction, mainly the structure by using two different kinds of reflective elements for each mirror, or using the same kind of reflective element but with different reflectance. It is simple to combine one reflective element with a cleaved fiber as the other mirror. The other is asymmetric in the radial direction. An example is the asymmetric IFFPI structure based on the off-axis fusion splicing. We call them axially asymmetric IFFPI sensor and radially asymmetric IFFPI sensor, respectively.

In the last section, two examples of the axially asymmetric IFFPI structures were introduced. One is the IFFPI structure with an air gap and a cleaved fiber end face as the mirrors [27]. The other is by using the splicing point between two different optical fibers [29], and a cleaved fiber end face as the mirrors. Such asymmetric IFFPI structures are simple to fabricate. As one of the mirrors is fixed by using the Fresnel reflection, it is easy to optimize the interference fringe contrast by designing the structure and making the reflection of the other mirror similar to that of the Fresnel reflection.

Axially asymmetric IFFPI structure based on a pair of FBGs with different reflectance was developed by Yongming Hu and coworkers [32]. Interference between multiple reflections from two strong FBGs was used to enhance the sensitivity. By using a path-matched Michelson interferometer, the sensitivity of the system was enhanced by more than 19 dB when the light was reflected nine times between the FBG mirrors.

Radially asymmetric IFFPI structure often has a lateral offset fabricated by using the fusion splicing method [33]. The fabrication process is rather simple, only by cleaving and fusion splicing. A large lateral offset of 62.5 μm was introduced during splicing two sections of SMFs. The reflections from the splicing point and the Fresnel reflection of the cleaved end show interference with each other. The interference visibility can be as high as 15 dB. This kind of radially asymmetric IFFPI sensor was used to measure high temperatures up to 1200°C with a sensitivity of 41 nm/°C. Experimental results also show that there was no degradation of the sensor performance during putting the sensor in a 1000°C environment for 160 min.

2.1.2 Extrinsic FFPI Sensors

Although the IFFPI structure has the advantages of high mechanical performance and easy fabrication, two drawbacks limit their applications. One is the cross sensitivity to temperature, which makes the test of all the other parameters not accurate when there are evident temperature variations. The other is the fact that there is no access for biochemical samples to go into the FP cavity. Thus, the IFFPI sensors can only use the end surface to perform the biochemical sensing, which is based on the relative intensity detection and may limit the sensitivity.

The initial motivation of introducing the extrinsic FFPI (EFFPI) structure is to solve these problems. EFFPI structures are often sandwiched with an air cavity between two reflective surfaces. The air cavity is important to reduce temperature sensitivity, or even to make temperature-insensitive sensors. The air cavity also offers great potential for biochemical samples to be located in the EFFPI structure, enabling the efficient interaction between samples and light and enhancing the detection sensitivity.

The performance of EFFPIs is strongly dependent on the cavity length and propagation loss inside it [34]. A strong loss of the cavity would enlarge the difference between the reflected intensities from the first and second reflective surface, and further degrade the interference fringe contrast. The loss of the EFFPI depends on the numerical aperture (NA) of the lead-in fiber, the air cavity length, the roughness, reflectance, and shape of the reflective surfaces. The NA of the lead-in fiber determines the divergent angle of light emitted out of the fiber. When cleaved optical fibers with flat surfaces are used to form an EFFPI, using lead-in fiber with lower NA should be better to reduce the propagation loss of light inside the cavity.

The cavity length of less than several hundreds of micrometers was often used to reduce the loss and obtain a fringe contrast better than 10 dB for high-performance sensing. Another way to enhance the performance is to use the Fizeau interferometer [35], which employs two surfaces with quite different reflectances with the far-end surface having a much higher reflectance. The high reflectance of the far-end surface can compensate the cavity loss and finally obtain a comparable reflected intensity from each surface. In this case, the cavity length can be extended up to several millimeters, greatly enhancing the multiplexing capability of the EFFPIs.

Up to now, the EFFPIs developed are mostly based on the two-beam interference by using the simple cleaved fiber end face. However, in some applications such as cavity quantum electrodynamics, high-finesse cavities are preferred. In these ultimate applications, flat-surface-based EFFPIs are not usable. Hunger et al. [36] fabricated high-finesse fiber FP cavities with concave, ultralow-roughness mirrors on the fiber end by using CO_2 laser pulses. According to the beam propagation theory in optical cavities, concave mirrors are very helpful to construct a stable cavity with ultralow loss. After coating, the cavity can have a finesse as high as 10^5.

On the other hand, the cavity length is often used as a sensing parameter to detect strain, pressure, or acoustics and needs to be controlled or precisely measured [37]. This part will be introduced in the interrogation and multiplexing section.

2.1.2.1 Capillary-Aligned EFFPI One of the most classic EFFPI structures is obtained by aligning the two cleaved fiber end faces by a glass capillary. In 1991, Murphy et al. [38] proposed this structure, as shown in Figure 2.7. A SMF was used as both the input and the output fiber. Its cleaved end face was one reflector. An MMF was used as another reflector that formed the EFFPI structure. The two fibers were aligned by a silica capillary. The Fresnel reflections were about 4% so that the first reflection dominates the interference and the subsequent reflections can be neglected. The interference fringes were mainly determined by the two-beam interference model, as discussed

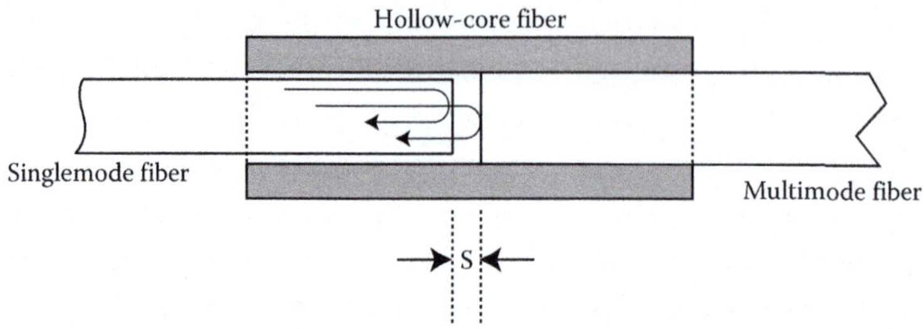

Figure 2.7 Capillary-aligned EFFPI structure.

in Chapter 1. The far end of the MMF was shattered so that the influence on the low-finesse cavity can be reduced. In this structure, the two fibers were easy to be moved in the silica capillary to adjust the air cavity length. Since then, this kind of simple EFFPI structure has been widely used, especially for high-pressure sensing by using the silica capillary as the sensitive element and the air cavity as the detection device. The silica capillary can withstand the high pressure, thanks to its relatively large thickness.

2.1.2.2 EFFPI Based on Fiber End and Diaphragm In order to enhance the sensitivity of pressure sensors, another EFFPI structure was proposed and also widely used. The EFFPI structure was formed by a cleaved fiber end face and a diaphragm. A general theory for diaphragm-based EFFPI can be found in Reference 39.

This kind of structure can be fabricated by the chemical etching method [40]. The fabrication process is shown in Figure 2.8. A section of MMF was first fusion spliced with a SMF, and the MMF was cleaved with a residual length of 40 μm similar to that of the final air cavity length. Then the MMF was chemically etched by HF acid. The etching rate of the MMF core is much faster, about 10 times than that of the pure silica cladding, enabling the selective etching of the MMF core. The etching process was stopped when the SMF core was exposed to the solution. The air cavity can also be formed by fusion splicing and precisely cleaving a section of hollow-core fiber. Then the SMF with an air hole at the tip was spliced to another section of SMF. The second SMF was cleaved and then polished so that the SMF length was controlled down to 3–5 μm, acting as the pressure-sensitive diaphragm. The polishing process was very important for the sensor performance.

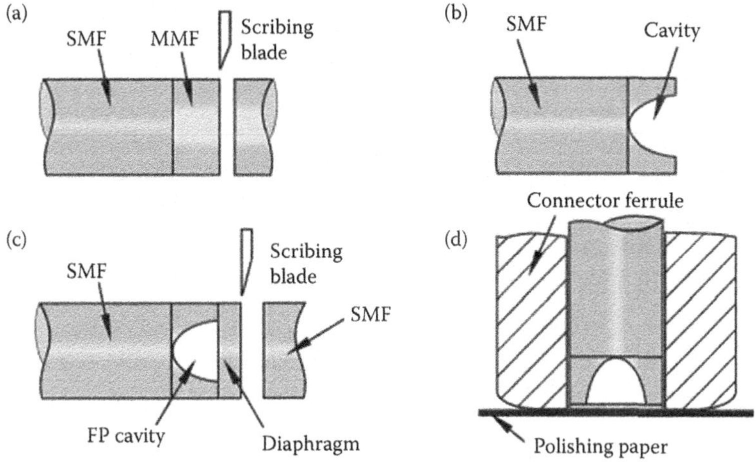

Figure 2.8 Fabrication of EFFPI with a diaphragm. (a) Fusion splicing an MMF to SMF and cleaving the MMF, (b) chemical etching the MMF, (c) fusion splicing a SMF to the etched MMF and cleaving the SMF, and (d) polishing the SMF to the desired thickness of the diaphragm.

Another simpler method is by using a fused silica ferrule with a V-shaped hole at one end [41]. A cleaved SMF was inserted into the silica ferrule, and a fused silica diaphragm was attached to the silica ferrule by heating fusion bonding. An EFFPI structure was formed at the V-shaped end of the silica ferrule. The V-shaped hole is important for enhancing the pressure sensitivity as it introduces a large effective diameter of the diaphragm. The thickness of the diaphragm is often optimized for the pressure detection with a certain range. A thick diaphragm is better for large pressure sensing, while a thin diaphragm is so for small pressure sensing. A similar structure was employed for the detection of the refractive index of a glass substrate. The fringe contrast of the two-beam interference from one cleaved fiber end face and the sample glass surface was used for the detection. Four different glass materials, that is, SiO_2, BK7, SF10, and SF11, were measured [42]. The standard deviation of the refractive index measurement for 20 times was less than 0.003. The method is simple, cost effective, easy to fabricate, and also easy to use. It was also used for small displacement measurement, with a resolution of 16 nm at an initial distance of 30 μm [43].

2.1.2.3 EFFPI Based on Microfabrication Technologies Nowadays, microfabrication technologies have been developed greatly, including chemical etching, excimer laser micromachining, femtosecond laser micromachining, and FIB milling, just to name a few. These technologies were widely used for the fabrication of various microelectromechanical systems (MEMS) and optical waveguides. The development of fiber-optic sensors also benefits a lot from these technologies, which make available EFFPIs with small size and high stability, as compared with ferrule aligned EFFPIs. The detailed description of the fabrication technologies can be found in Chapter 3. Here, we are going to describe the characteristics of the fabricated EFFPI structures.

Chemical etching methods have been used for shaping the microstructure of optical fibers for a long time. The fiber-optic probes with sharp tapers at nanometer scale were chemically etched for near-field optical microscopy [44,45], while the etching occurred at the interface between the etching solution and air due to the surface tension effect. Different from the fabrication of nanofiber tapers, the EFFPI structure can be fabricated by another chemical etching mechanism, using the etching rate difference between the fiber core and cladding [46,47], as well as the fusion splicing process.

Usually, HF acid, sometimes buffered HF acid (HF mixed with NH_4F), was used for the etching. In order to obtain a higher sensing performance, EFFPI can also be fabricated by chemically etching Er-doped fibers with mixed hydrochloric (HCl) and HF acid and fusion splicing, as shown in Figure 2.9. The sensor performance is greatly improved by the chemical reaction between HCl acid and doped Er_2O_3. A maximum fringe contrast of ~24 dB is obtained [47], comparable to that of MEFPI sensors fabricated by excimer lasers. The MEFPI sensor has high mechanical strength as the etching rate difference between fiber core and cladding is enlarged. This kind of sensor is insensitive to temperature while highly sensitive to strain, with sensitivities of ~0.65 pm/°C and ~3.15 pm/$\mu\varepsilon$, respectively.

It is worth noting that the insensitivity to temperature is a universal advantage of EFFPI sensors, partially due to the low thermo-expansion coefficient of air and the self-compensation of the structure. The self-compensation means that, as the temperature increases, the cavity length remains almost the same, because the cladding part of the EFFPI tends to stretch the structure and increase the cavity length, while the fiber core part tends to be expanded into the cavity and make the cavity length shorter.

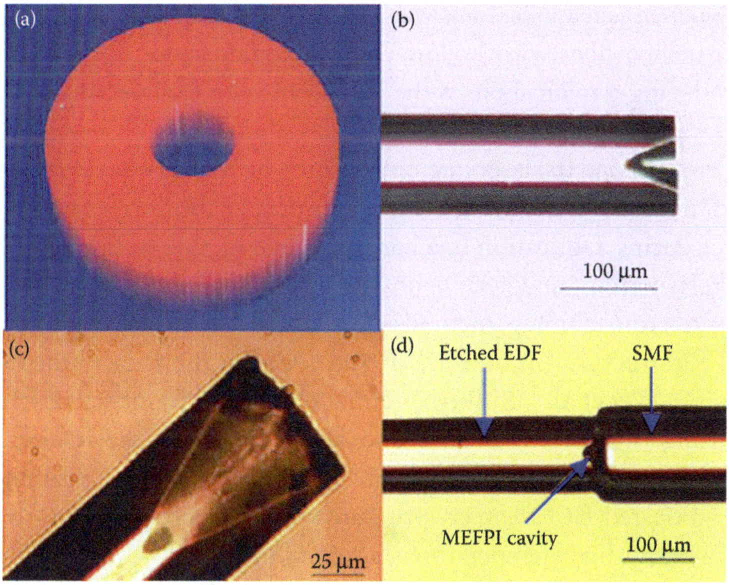

Figure 2.9 (a–c) Microscopic images of the etched optical fiber and (d) EFFPI sensor by chemical etching.

The main advantage of chemically etched EFFPIs is the simple and low-cost fabrication process. However, the final sensing performance is strongly dependent on the fusion splicing process, as the electric charge intensity and period need to be optimized through the fusion splicing program in order to melt the surface for two-beam interference. The fusion splicing process lacks repeatability and may also be influenced by the residual etching solution. Another disadvantage is that the chemically etched surface is cone-shaped, which is not preferred for high-quality EFFPIs.

The laser micromachining technologies were employed for fabricating EFFPI structures, by using either the 157-nm excimer laser or femtosecond laser micromachining systems [48–50]. High-quality surfaces can be fabricated and the interference fringes were quite good and the fringe contrast can be as high as 26 dB. The smoothness of the fabricated surface is very good by using a mask-based fabrication technique and using a laser spot with uniform intensity distribution. By combining the fabricated micro-hole with a diaphragm, high-performance pressure sensors can be formed. This kind of EFFPI sensors is especially useful for strain and pressure sensing at high-temperature environments [49]. In addition, refractive index sensing for biochemical applications was also performed by fusion splicing the laser-etched fiber with hollow-core photonic crystal fiber (PCF) or introducing a micro-hole at the end as an inlet for injecting aqueous samples [50].

Femtosecond (fs) laser micromachining technology has been widely used in recent years due to its capability of cold ablation. Very low heat effect during fabrication is a big advantage of this technology. Now microstructures, and even nanostructures, have been fabricated by fs laser micromachining methods. In 2007, Rao et al. [51] fabricated EFFPI structures in both SMF and PCF by the fs laser micromachining. Wei et al. [52] further enhanced the performance of EFFPI by using a Coherent fs laser and also monitoring the interference spectra of the fabricated device in real time. The fringe visibility of the fabricated EFFPI microstructure was enhanced up to 14 dB and was used for high-temperature sensing of up to 1100°C. Good linearity was observed for the high-temperature sensing.

EFFPI structures are especially useful for physical parameter sensing in harsh environments. However, after ablation by the fs laser, the mechanical strength of EFFPI was not as good as that of the fiber before the ablation. The only exception is the self-enclosed all-fiber in-line EFFPI structure [53]; only the fiber core part is ablated and the fiber cladding can withstand relatively large strain and also dynamic strain.

The most promising aspect of the micromachined EFFPI is for biochemical applications, as the micronotch is not only a fine micro interferometer, but also an open sampling channel for biochemical reagents. When the refractive index of the medium within the cavity was changed, the interference fringes of the reflective spectra would shift [54].

Besides the chemical etching and laser micromachining methods, FIB milling [55] was also used for the fabrication of EFFPI. By using the FIB milling technology, a compact open-hole EFFPI refractive index sensor was fabricated and integrated with a microfluidic channel. The thermal effect makes the reflective surface smooth; however, it may sometimes introduce deformations of the fabricated device and degrade the fringe visibility and sensing performance. Because of the strong thermal effect, it is very difficult to use the CO_2 laser pulses to fabricated precise EFFPI microstructures in optical fibers.

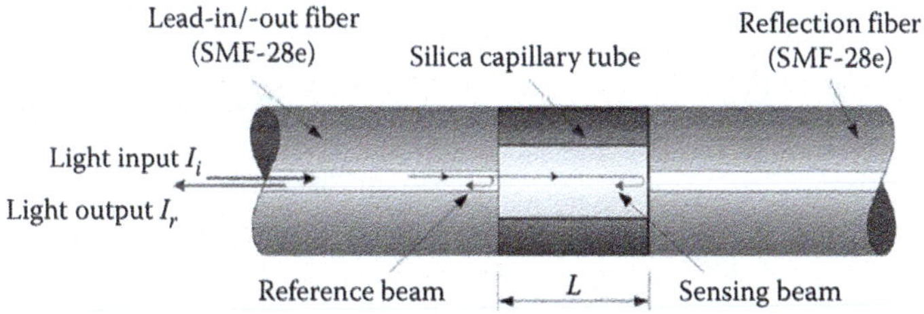

Figure 2.10 EFFPI structure formed by splicing a capillary in between two SMFs.

2.1.2.4 EFFPI Based on Capillary or Hollow Fiber Glass capillaries with similar dimensions to that of optical fibers, named hollow-core fibers, can be used for the formation of EFFPIs, by fusion splicing a short section of capillary between two SMFs [56], as shown in Figure 2.10. The fabrication process is simple and includes only cleaving and fusion splicing. No expensive micromachining systems are required and it is cost effective. The fringe visibility can achieve as high as 20 dB when the cavity length is around 60 μm. The cavity length of longer than 1 mm was also demonstrated, with a relatively lower visibility of about 8 dB, which indicates the great capability of spatial frequency multiplexing. There might be a problem on the automatic alignment during the fusion splicing. As the alignment control of a splicer was usually based on the feedback of the imaging of the fiber core, the hollow-core fiber or capillary with air in the core area may require manual alignment during the fusion splicing.

Besides by the use of micromachining technologies, open-hole EFFPIs can also be formed by aligning two SMFs through an open-hole capillary with diameter at millimeter scale. The open hole can be easily ground by conventional mechanical tools. The capillary and SMFs can be bonded by CO_2 laser welding to enhance the stability of the device [57]. This is a low-cost way to fabricate a sensitive EFFPI sensor, which is capable of measuring the biochemical parameters.

2.1.2.5 EFFPI Based on Bubble Microbubbles or microspheres have been used for developing high Q factor optical resonators, which have supported whispering gallery modes for decades. Thanks to the development of high-end fusion splicers, microbubbles can be fabricated with good controllability [58], as shown in Figure 2.11. A unique rectangle air-bubble-based EFFPI was reported recently by fusion

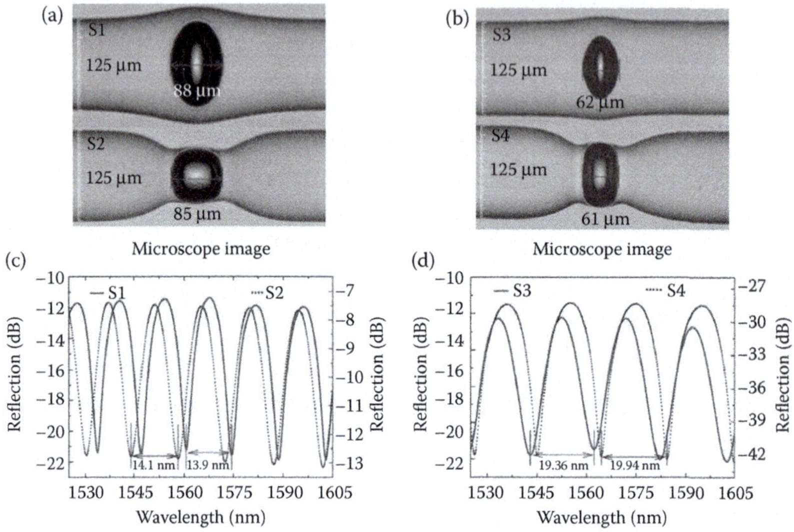

Figure 2.11 (a, b) EFFPI structures based on microbubble formed by fusion splicing and (c, d) corresponding interference spectra. The parameters for fusion splicing were different between (a) and (b).

splicing two cleaved SMFs and tapering the splicing point using the same fusion splicer. The surface was very smooth as the fabrication process occurs after melting of the material and the wall thickness can be controlled as thin as 1 µm. Due to this unique microstructure, the strain sensitivity was as high as 43 pm/µε, and the temperature cross sensitivity was low, 0.046 µε/°C.

2.1.2.6 EFFPI Based on Reflective Film Most of the previous EFFPI structures were formed by two surfaces with low reflectance, often determined by the Fresnel reflection formula. In these low-finesse cavities, a two-beam interference model can be used to describe it, as the intensities of higher order reflections are much lower than the first reflection from each reflective surface. Usually, the reflection spectra were detected in the case of low-finesse FP cavities. If we look at the transmission of the low-finesse cavities, the intensity of the first transmission is much stronger than the second or even higher order transmission, so that the fringe visibility is very low due to the strong dc signal.

High-finesse cavities were not preferred in the field of FFP sensors owing to their relatively complicated fabrication process. However,

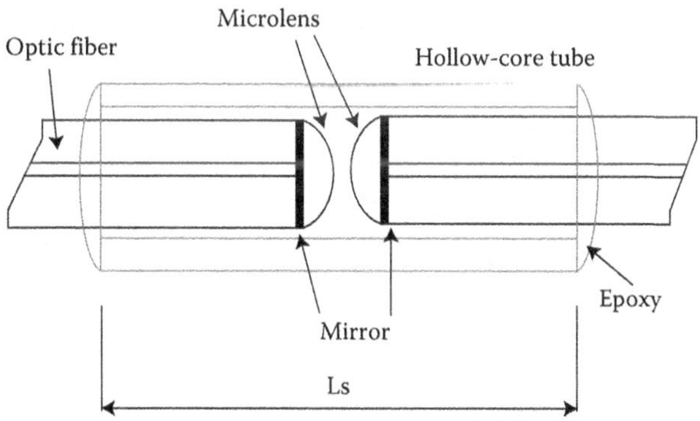

Figure 2.12 High-finesse EFFPI based on coated SMFs and microlens pairs.

high-finesse cavities correspond to a much narrower linewidth of resonant peaks, compared to the interference fringes in low-finesse cavities.

High spectral resolution leads to high resolution for sensing applications. In 2008, Jiang et al. developed a high-finesse microlens-based EFFPI by curing epoxy droplets on the fiber ends, as shown in Figure 2.12. This method is simple and cost effective [59].

The manufacturing procedure is described as follows. First, two SMFs were cleaved, polished, and coated with highly reflective films. The reflectance was found to be about 96%. Second, another bare optical fiber was cleaved and dipped into the epoxy adhesive (NOA61, Norland) with a depth of 1 mm. When the fiber was withdrawn from the liquid epoxy, a small droplet of epoxy was formed at the end of the fiber. This epoxy added fiber and the coated fiber were fixed at each side of the fusion splicer. The two fibers were pushed together and then separated, resulting in a small epoxy droplet being attached on the fiber end with a mirror. Then the mirror-coated fiber with a small epoxy droplet was put downward letting gravity make the curve of the lens symmetrical, and was finally cured using a UV light for 10 min. To form the highfinesse EFFPI, two mirror-coated SMFs with epoxy microlens were inserted into a capillary for alignment.

This high-finesse EFFPI structure has two differences from other EFFPIs. One is that highly reflective films were coated on the fiber ends, reducing the transmission loss of the cavity. The other is using the epoxy-based microlens, which reduces the propagation loss of multiple reflected beams in the cavity. Due to the high reflectance of the coated mirror, the intensities of the transmitted beams are comparable and multi-beam interference, rather than two-beam interference, occurs in the cavity. Indicated by the transmission spectrum of the high-finesse EFFPI, the FSR is 14.5 nm and the finesse of this kind of EFFPI is around 37 and 78, which is not as high as expected from the high reflectance of 96% because of the divergent angle of light emitted from each fiber end. But the finesse is much higher than the EFFPIs with Fresnel reflections. The high-finesse EFFPI has high sensitivities for strain and temperature sensing, with resolution of 0.0625 με and 0.025°C, respectively.

Jiang and his coworkers [60] further fabricated a high-finesse EFFPI for magnetic field sensing. The fabrication process was the same as described above, except one TbDyFe rod was used at one end instead of SMF. The light was incident from the lead-in SMF and reflected from the coated film on the TbDyFe rod. The reflected light was detected as the TbDyFe rod was not transparent. The length of the TbDyFe rod was changed with the ambient dc magnetic field, introducing a cavity length change of the EFFPI. The wavelength shift was measured with the magnetic field. The sensor had a high sensitivity of 1510 nm/mT, with a magnetic resolution of 25 nT. Again it is proved that the high spectral resolution of the resonant wavelength peaks is beneficial for high-resolution sensing.

2.1.2.7 EFFPI Based on Sensitive Film EFFPI fabricated with clean optical fiber end face can measure physical parameters such as strain, pressure, temperature, or even refractive index-based biochemical parameters. However, they are not able to perform humidity detection, gas sensing, or biosensing. Sensitive films have been widely integrated with various fiber-optic structures in order to strengthen the capability of fiber-optic sensing. There has been great progress, and considerable contribution has been made by researchers from a variety of fields. Here, we only introduce several examples.

In 2009, Fan and coworkers [61] developed an EFFPI gas sensor by coating a silver film and versatile vapor-sensitive polymer layer on a cleaved fiber end face. The interference occurs between the reflected light from the silver–polymer interface and the polymer–air interface. Any polymer materials, regardless of the refractive indices, can be used for the proposed microstructure. Two kinds of polymers, polyethylene glycol (PEG) 400 (RI = 1.465–1.469) and Norland Optical Adhesive (NOA) 81 (RI = 1.53–1.56), were used as examples for gas sensing and showed drastically different sensor response to hexanol, methanol, and acetone. The thickness of the polymer layer was about 30 μm. The sensitivity for methanol vapor was 3.5 and 0.1 pm/ppm for PEG 400 and NOA 81, respectively, with a detection limit on the order of 1–10 ppm. The introduction of the silver film greatly enhanced the fringe visibility, as it has a very low refractive index, $n = 0.3 + 11i$. Also it enables the versatile usage of various polymer materials.

By using the SU-8 photoresist as the coating material, EFFPI was fabricated based on the film-on-fiber structure [62]. The SU-8 film was coated directly on the cleaved fiber end face. The thickness of SU-8 film was about 30 μm. The fringe contrast was measured from the reflective spectra as a function of the refractive index of the medium solution. The fringe visibility reached ~25 dB when the device was immersed into water. An average sensitivity of 205 dB/RI was achieved in the range between 1.314 and 1.365. The resolution can be estimated to be 5×10^{-6} as the relative intensity resolution of the optical spectrum analyzer can reach 0.001 dB near 1550 nm. Temperature insensitivity and long-term stability for RI sensing were confirmed experimentally. In another work, the authors also investigated the wavelength shift of the interference fringe of the SU-8-coated EFFPI as a function of temperature [63]. The sensitivity is about 200 pm/°C. The standard deviation of the peak wavelength was about ±300 pm over 60 min when the temperature was fixed at 50°C.

2.2 Multimode FFPI Sensors

Most of the developed FFPIs were based on SMFs, due to the low cost and good preservation of the single transverse mode. The latter is good for maintaining the high performance of FFPI sensors. However, multimode fibers (MMF) were sometimes employed for forming FFPI structures with special characteristics. The unique characteristic of MMF is that the core diameter, often 62.5, 50, or 105 μm, is much larger than that of the SMF, 8–10 μm for SMF at the wavelength of 1550 nm.

One example of multimode FFP is by using the MMF as an efficient light collecting element [64]. The structure of the multimode

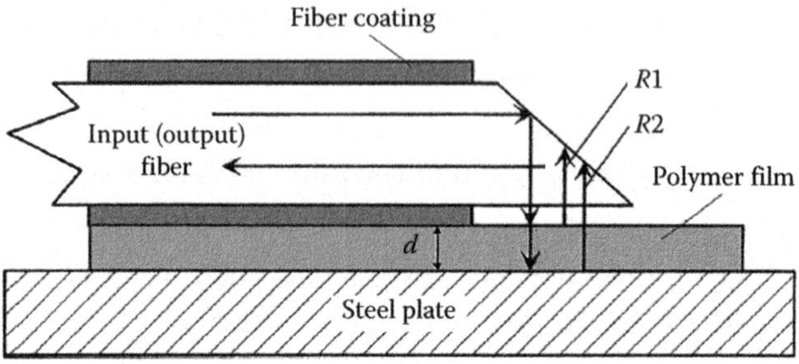

Figure 2.13 FFPI using MMF as the input and output fiber. (Reprinted from Jiang, M. and Gerhard, E. 2001. A simple strain sensor using a thin film as a low-finesse fiber-optic Fabry–Perot interferometer. *Sensors and Actuators A: Physical*, 88(1), 41–46, Copyright 2001, with permission from Elsevier.)

FFPI is shown in Figure 2.13. The FP cavity was formed by the polymer film. The MMF polished at 45° was used as a lead-in and read-out channel for the light. The polymer film was attached to a steel plate for constructing an attachable strain sensor. The film was made from polyurethane, which was composed of a mixture of two-phase structure domains called hard and soft segments, which provided the high tensile strength and flexibility to adjust the physical properties of the film. When strain was added on the device, the thickness of the polymer film would change, which can be determined by the shift of interference fringes. Wu and Solgaard [65] investigated the short-cavity-length EFFPI pressure sensor, which employed MMF as the input and output fiber and the microfabricated diaphragm based on the silicon-on-insulator (SOI) wafer, as shown in Figure 2.13. They also discussed in detail the effects including the mode averaging, phase uncertainty, amplitude reduction, and spectral modal noise.

There were also other kinds of multimode FFPIs where MMF was not only used for the input and output of light, but was engaged directly in forming the structure. Besides its core diameter being larger than SMF, MMF often has higher NA than SMF, 0.12 or 0.14. It is easy to find commercial MMF with NA between 0.2 and 0.4, or even higher, designed for high-efficient collection of light in the field of imaging and spectroscopy. High NA MMF is useful for fabricating FFPIs, thanks to its high doping of Ge, which makes a higher etching rate difference between fiber core and cladding. The dissociation energy of the Si-O bond, 799.6 kJ/mol, is larger than that of the Ge-O bond, 660.3 kJ/mol. The fiber cladding was often composed of SiO_2, while the fiber core was doped with high doping concentration of GeO_2. The latter can be etched by HF acid with a higher etching rate than that of SiO_2. The large core diameter is also helpful for forming a relatively flat reflective surface compared with a small core SMF.

By chemically etching a graded-index MMF with a core diameter of 105 μm and NA of 0.3, a notch was obtained at the fiber end face with a depth of 25 μm when etched for 90 s with 40% HF acid [66]. An EFFPI with a cavity length of 17.5 μm was fabricated after fusion splicing with another cleaved fiber and the fringe visibility was 22 dB. Usually, the cavity length of FFPI was shorter than the depth of the micromachined notch due to the partial overlap of the two fibers and compression of the cavity length during the splicing. This kind of multimode EFFPI sensor preserves the advantages of other EFFPIs like low temperature cross sensitivity.

In the work presented above, the MMF was used as a reflective mirror only, similar to the EFFPI pressure sensors with other kinds of diaphragm [67] for biomedical applications [68]. Another motivation of using multimode optical fiber in FFPIs is enhancing the sensitivity for pressure sensing by using the large-diameter fibers as the diaphragm.

Ma et al. [69] proposed an IFFPI based on the SMS structure. Different from the modal interference detection and from the transmission spectra, the authors investigated the multimode excitation-induced phase shifts in the reflected spectra. Further, they fabricated SMS IFFPIs with graded-index MMF [70]. The insertion loss was reduced and the multiplexing capability was greatly enhanced by the refocusing effect of light propagation in the graded-index MMFs.

2.3 Hybrid FFPI Sensors

Up to now, we introduced extrinsic and intrinsic FFPIs based on either SMFs or MMFs. IFFPIs have the advantages of high-temperature sensitivity and high mechanical strength, while EFFPIs have low-temperature cross sensitivity and are especially useful for temperature-insensitive sensing of other kinds of parameters such as pressure, strain, etc. SMF-based FFPIs have the advantages of good transverse mode distribution and no cross talk from multimode excitation, while MMFs are much easier for alignment and easy-to-use for high-efficiency light coupling during the fabrication of FFPIs. Therefore, researchers in this field have made further efforts on making full use of all these advantages by fabricating hybrid FFPIs.

2.3.1 Intrinsic–Extrinsic Hybrid FFPIs

As described in Section 2.1.2, the in-line EFFPI based on an air cavity was first used for strain sensing. By simply cleaving one end of the air-cavity-based EFFPI, an intrinsic–extrinsic hybrid FFPI was fabricated, as shown in Figure 2.14. The end face can be exposed to aqueous solutions to measure the refractive index [71], without the complicated process of fabricating the microchannels for sampling. It has the advantages of a large measurement dynamic range for refractive index sensing (Figure 2.14).

Another similar structure, by fusion splicing a short section of hollow-core fiber and SMF to the lead-in PCF [72], was developed for high-temperature sensing. By fusion splicing a SMF with a large-diameter capillary and a small-diameter capillary in sequence, an intrinsic–extrinsic hybrid FFPI was developed for gas sensing [73]. The air gap based on the large-diameter capillary formed an EFFPI, while the small-diameter capillary formed an IFFPI and also served as the sampling channel for gas.

Another advantage of using the intrinsic–extrinsic hybrid FFPI is enhancing the fringe visibility of the reflected spectra by three-beam interference, instead of two-beam interference. The schematic diagram of the three-beam interference model is shown in Figure 2.15.

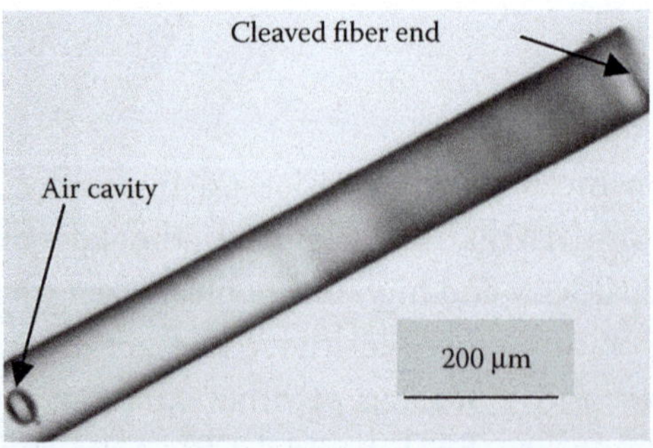

Figure 2.14 Intrinsic–extrinsic hybrid FFPI for refractive index sensing.

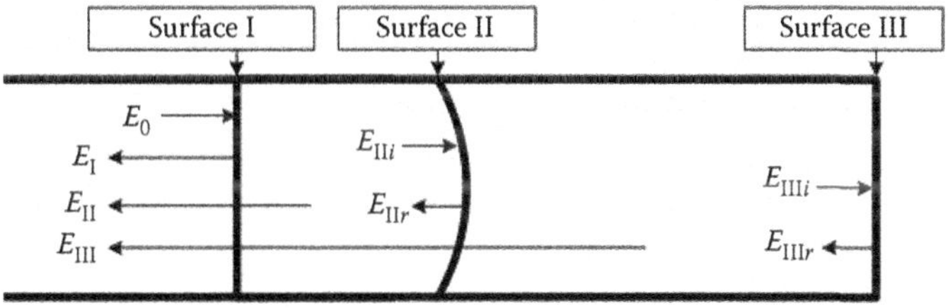

Figure 2.15 Schematic diagram of the three-beam interference model for the intrinsic–extrinsic hybrid FFPI.

The theoretical expressions were described in detail in Reference 74 and also in Chapter 1.

As is well-known, the fringe visibility of two-beam interference reaches its maximum value when the reflectance of the two FFPI surfaces equals, that is, $R_I = R_{II}$. R_I, R_{II}, and R_{III} are the reflectance of the reflective surfaces of the FFPI. If $R_I = R_{II}$ is not strictly satisfied, the fringe visibility will not be very high.

Considering the three-beam interference, the fringe visibility becomes maximum when the reflectance of three reflective surfaces meets the following inequality:

$$-1 \le \frac{R_I + R_{II} - R_{III}}{2\sqrt{R_I R_{II}}} \le 1 \tag{2.3}$$

This makes it much easier for the three-surface-based hybrid FFPI sensor to obtain fringe visibility higher than the conventional two-surface FFP sensors, as shown by the gray curve in Figure 1.6. The fringe visibility can be as high as 35 dB [75]. The high fringe visibility is helpful to enhance the sensing resolution due to the higher spectral resolution, if the wavelength shift is detected, and also to enhance the sensitivity if the fringe visibility is detected as a function of the measurand.

By combining the singlemode–multimode hybrid FFPI structure, not only can the fabrication process be simplified by reducing the precision of alignment, but this is also useful for multi-parameter sensing. The use of graded-index MMF is also helpful for enhancing the fringe visibility of the reflected spectra of the FFPIs, which can be as high as 35 dB [76]. For refractive index sensing based on fringe visibility detection, the sensitivity of 160 dB/RIU can be achieved [74].

2.3.2 *Hybrid FFPI with Other Fiber-Optic Structures*

It is a good strategy for multi-parameter sensing to combine the FFPI with other types of fiber-optic structures. In this way, the advantages of different kinds of fiber-optic structures can be integrated and the capability can be enhanced while the size remains small. Rao et al. developed a hybrid structure by integrating a long-period fiber grating (LPFG) and an EFFPI together [77]. The reflective spectra of the EFFPI and the transmission spectra of the LPFG were detected at the same time for the multi-parameter sensing. High temperature and strain were measured simultaneously. Due to the fact that only pure silica fibers were used, temperature up to 650°C was measured. Strain was also measured at a temperature of 500°C. This work made use of both the reflective and transmission spectra of two different kinds of fiber-optic structures for sensing.

Frazao et al. [78] demonstrated multi-parameter sensing by using hybrid FFPI and a Michelson interferometer. Both temperature and strain were measured. The device was highly integrative by using a dual asymmetric core microstructured fiber. The two interferometers had different sensitivities for strain and temperature measurement. By using the matrix of the strain and temperature coefficients, multi-parameter sensing was achieved.

2.4 Microstructured FFPI Sensors

2.4.1 *Photonic Crystal Fiber FFPIs*

PCFs including index-guided PCF and photonic bandgap fiber as two main types, have been extensively investigated since they were invented in the 1990s [79,80]. As the technologies for the fabrication and characterization of PCFs, as well as the fusion splicing of PCF with SMFs, were developed, PCFs were used for enhancing the performance of FFPIs. PCF–FFPIs have been used for the measurement of various parameters, including high temperature, strain, pressure, and refractive index.

Fiber-optic sensors have the advantage of being capable of surviving in harsh environments such as those with high temperatures. FFPIs are even better for high-temperature sensing than the conventional FBGs. One example of the PCF–FFPIs was fabricated by fusion

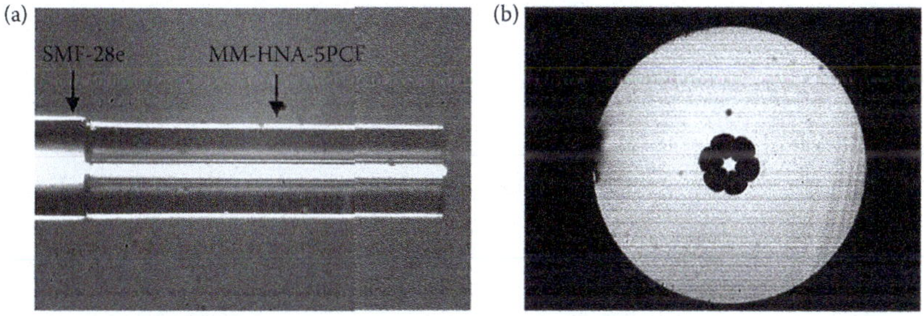

Figure 2.16 (a) FFPI fabricated by fusion splicing a short section of PCF with SMF and (b) the cross section of the PCF. (Reprinted from Zhu, T., Ke, T., Rao, Y., and Chiang, K. S. 2010. Fabry–Perot optical fiber tip sensor for high temperature measurement. *Optics Communications*, 283(19), 3683–3685, Copyright 2010, with permission from Elsevier.)

splicing a short section of solid-core PCF with a SMF [81], as shown in Figure 2.16a. The PCF has a solid core with a diameter of 5.1 μm, surrounded by six large holes with a diameter of 10.5 μm, as shown in Figure 2.16b. The interference occurred between the reflections from the SMF–PCF interface and the end face of the PFC core. The fringe visibility was about 12 dB, which was high enough for harsh environment sensing. The PCF–FFPI can measure temperature as high as 1200°C due to its pure silica structure. The repeatability of sensing was good. Another similar FFPI structure was fabricated by using a section of polarization-maintaining PCF and high-temperature sensing was also demonstrated [82].

A section of hollow-core PCF was fusion spliced in between two SMFs to form an EFFPI [83], as shown in Figure 2.17. The cavity length can be as long as several centimeters with good fringe visibility due to the low transmission loss of PCF compared with hollow-core fibers (or capillaries), which is important to enhance spatial frequency

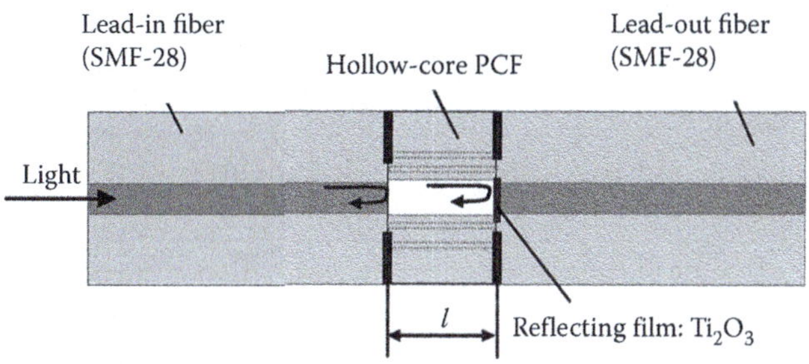

Figure 2.17 EFFPI based on fusion splicing a section of PCF in between two SMFs.

multiplexing capability. The fringe contrast can be greatly enhanced by coating the end face of SMFs and then fusion splicing with PCF. Based on this simple PCF–EFFPI structure, temperature-insensitive strain sensing was performed.

Other kinds of PCFs were also used to form similar PCF–EFFPI structure, with different sensing performance [84,85]. Similar structures were employed for pressure sensing based on the lateral deformation of the PCF [86].

PCFs can also be used as a hollow-core fiber for fabricating the intrinsic–extrinsic hybrid EFFPI, similar as a hollow-core capillary. In 2008, Rao et al. [87] developed a PCF–FFPI structure for refractive index sensing. A hollow-core PCF was fusion spliced with a SMF and then cleaved with a PCF length of 2.3 mm. The fusion splicing procedures and parameters for PCFs were different from the conventional ones between two SMFs. Then another SMF was fusion spliced to the PCF and then cleaved by a fs laser with a SMF length of 20 μm. The PCF part was an EFFPI, while the short SMF cap was an IFFPI. As calibration, the fringe visibility was measured as a function of the refractive index. Experimental results indicated the fringe visibility of the PCF hybrid FFPI was insensitive to temperature. Another similar intrinsic–extrinsic hybrid PCF–FFPI structure was fabricated by changing the type of PCF, the length of PCF, and SMF [88]. The fringe visibility was enhanced up to about 30 dB when the hybrid PCF–FFPI was put in air, and was about 8 dB when putting the device in water.

Due to the multi-hole structure of PCFs, the fusion splicing parameters should be specifically set up for fabricating the PCF–FFPI structure. There was another way to fabricate a PCF–FFPI. As described in Reference 89, a film was formed by fully collapsing the air hole of a hollow-core PCF, after fusion splicing the PCF with a SMF. Further, they fabricated a PCF–FFPI by forming two films through collapsing the air holes at each end of the PCF [90].

The most promising aspect of PCF for fiber-optic sensing is the multi-hole structure, which is excellent for biochemical sampling and also for efficient light–matter interaction. The label-free biochemical detection is usually performed by refractive index measurement. An ultrahigh sensitivity, of 38,000 nm/RIU, for fiber-optic refractive index sensing [91] was obtained by detecting the wavelength shift

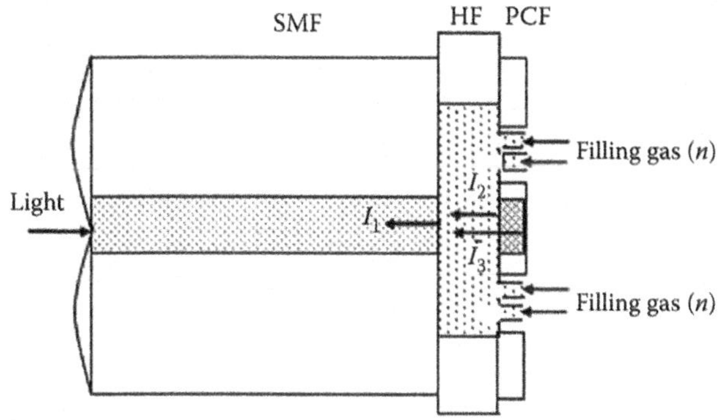

Figure 2.18 PCF–FFPI with liquid sampling channels.

from the mode coupling in PCF and by filling a single hole of PCF. The detection limit of refractive index was on the order of 10^{-7}.

Deng et al. [92] fabricated a PCF–FFPI by fusion splicing a short section of hollow-core fiber and PCF in sequence with a SMF. The structure is shown in Figure 2.18. The solid core of the PCF served as a reflection surface, and the air holes were sampling channels for external gas under test to enter into the hollow fiber chamber. A low fusion power and a short fusion time were used during the fusion splicing of PCF, in order to avoid the collapse of the air holes. The sensitivity of wavelength shift versus refractive index was 1639 nm/RIU, which was higher than many other kinds of fiber-optic refractometers.

2.4.2 Microstructured Optical Fiber FFPIs

Besides the PCFs, there are other kinds of MOF including suspended-core fiber (SCF) and polarization-maintaining fiber (PMF). FFPIs based on SCF and PMF were also developed [93–95].

SCF–FFPIs were fabricated by fusion splicing a short section of SCF with SMF and used for strain, temperature, and refractive index sensing. One SCF had three holes with a diameter of 20 μm and a core diameter of 3.2 μm. Another SCF had four holes with a diameter of 43 μm and a core diameter of 5 μm. The lengths of the two SCFs were 1 and 0.84 mm, respectively, for the three- and four-hole SCFs. In order to exert strain on the FFPI, the cleaved end of the SCFs was spliced with a hollow-core PCF. The strain and temperature sensing performance of the SCF–FFPIs had no big difference from FFPIs

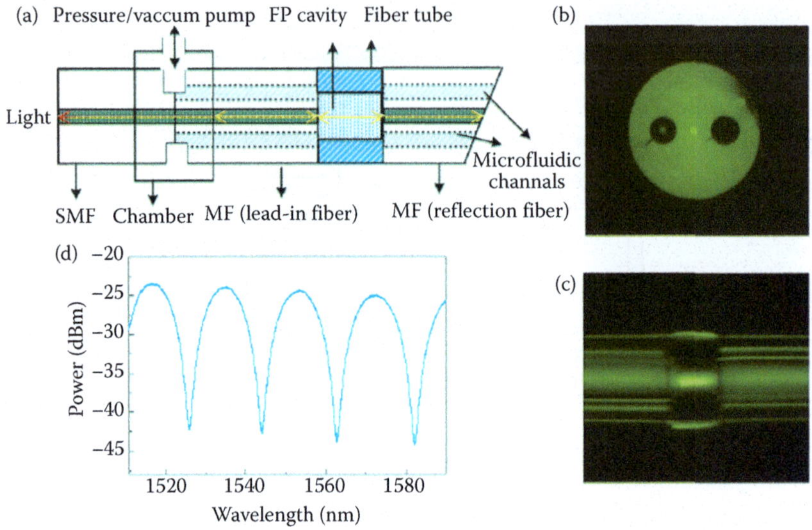

Figure 2.19 PMF–FFPI structure and its reflective spectrum.

using other types of optical fibers. However, it was good for refractive index sensing, thanks to the large holes around the suspended core.

Polarization maintain fibers (PMF) were also used for fabricating FFPIs based on inscription of a pair of FBGs [96]. The relatively high reflectance of the FBGs enabled a fine spectra with FSR of 40 pm and a linewidth of each fringe of 1 pm, which was helpful for high-resolution sensing.

PMF, as another main kind of microstructured fibers, can be employed for the microfluidic applications by using the two symmetric air holes in the fiber cladding [97,98]. The microfluidic PMF–FFPI structure is shown in Figure 2.19. The solid-core diameter was 6 μm and the diameter of the air holes is 26 μm with a center-to-center distance of 60 μm. By fusion splicing the PMF, a fiber tube, PMF in sequence with a SMF, FFPI was formed based on the reflections from the two ends of the fiber tube, which had an inner diameter of 75 μm for liquid sampling [97]. An inlet was introduced by etching the joint of the SMF and the PMF and the microfluidic flow was activated by an external pressure/vacuum pump. The far end of the PMF was cleaved at an angle to avoid reflections. The RI sensitivity of 1051 nm/RIU was achieved, with good repeatability and low cross sensitivity of temperature.

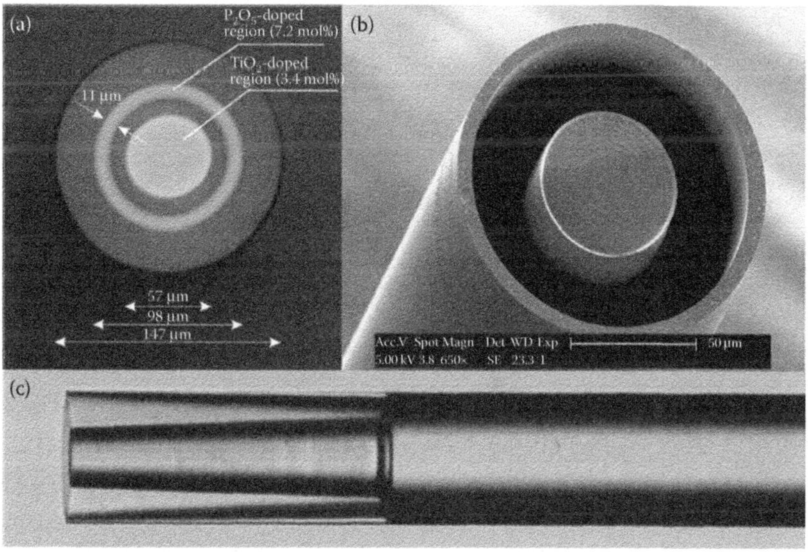

Figure 2.20 (a) Specialty optical fiber designed for fabricating FFPI sensors and (b) SEM and (c) microscopic images of the after etching.

There were FFPIs based on purpose-designed MOFs. One example is shown in Figure 2.20, with the central core region doped with TiO_2, a pure silica barrel ring, a P_2O_5-doped ring, and an outer cladding [99]. The P_2O_5-doped ring can be etched at a high rate, and the TiO_2-doped core has a low etching rate. Therefore, a flat core end face, as well as a thin cladding, can be fabricated after chemical etching, as shown in Figure 2.20b and c. After fusion splicing with a lead-in SMF, the MOF-FFPI with a short cavity length was formed (Figure 2.21). However, thanks to the thin, but relatively long

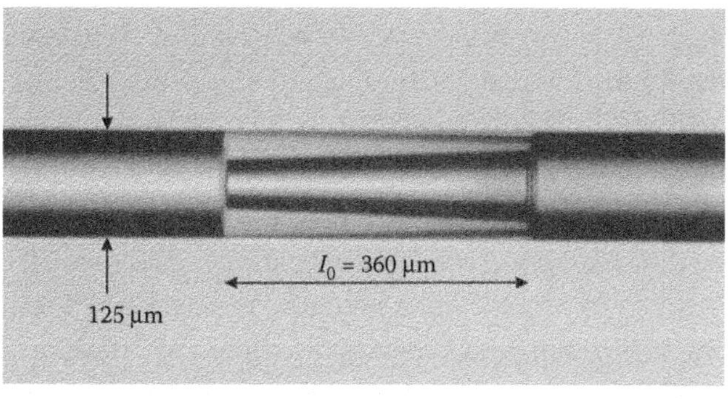

Figure 2.21 FFPI formed by the etched microstructured fiber.

(360 μm), etched cladding, a strain resolution of 0.5 με and low cross sensitivity of 0.055 με/°C were obtained.

2.4.3 Microfiber or Tapered FFPIs

FFPIs using optical microfibers or tapered optical fibers with a diameter of several micrometers are specially used for highly sensitive force sensing. Gong et al. [100] fabricated FFPIs with interference between a low-reflectance FBG and a Fresnel reflection fiber end face. The thin optical microfiber in the FP cavity served as a force-sensitive element. The wavelength-force sensitivity was 0.221 pm/μN, which was much higher than a similar structure with 125 μm-diameter SMF.

Besides the force-sensitive performance, the microfiber FFPI is also sensitive to refractive index, thanks to its strong evanescent wave penetrated into the medium. Zhang et al. [101] fabricated such a microfiber FFPI by heating and drawing an FBG in the middle. The two half FBGs acted as the reflective mirrors for the FFPI and the optical microfiber was used for refractive index sensing.

The FBG can be fabricated directly into optical microfibers by the 193-nm UV laser exposure, by either the phase mask or point-by-point writing technologies. FFPI based on a pair of microfiber FBG was fabricated, and sensing performance was investigated in detail by Guan and coworkers [102].

In addition to the advantages of force- and index-sensitivity, the FFPI on a fiber taper has a smaller size than the FFPIs based on common SMFs. Kou et al. [103] developed a miniaturized FFPI on a fiber taper tip. The fiber taper had a total length of 2.1 mm and the tip diameter was less than 1 μm. After coating a 150-nm-thick aluminum (Al) film, a microcavity was fabricated by FIB milling technology. The cavity was 4.4 μm long and 5 μm high. Then the Al film was removed by HF acid etching. High-temperature sensing was carried out. The SEM images of the FP cavity and the tip were shown in Figure 2.22.

Totally speaking, there have been various FFPI microstructures developed for different kinds of applications. Although it has now been developed for decades, the potential of novel FFPI structures is continuing to be explored, for example, the newly developed microbubble and PCF-based FFPI sensor, which pushed the sensitivity of the refractive index to a new level. We believe that new structures

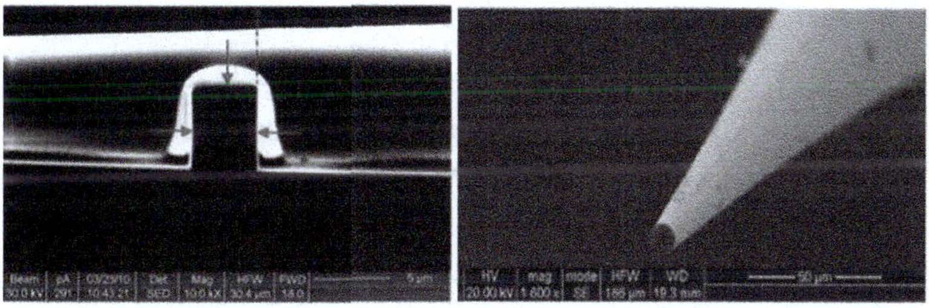

Figure 2.22 FFPI on fiber taper.

and new schemes will continue to enhance the performance of FFPI sensors in future.

References

1. Kawasaki, B. S., Hill, K. O., Johnson, D. C., and Fujii, Y. 1978. Narrow-band Bragg reflectors in optical fibers. *Optics Letters*, 3(2), 66–68.
2. Kersey, A. D., Davis, M. A., Patrick, H. J., LeBlanc, M., Koo, K. P., Askins, C. G., Putnam, M. A., and Friebele, E. J. 1997. Fiber grating sensors. *Journal of Lightwave Technology*, 15(8), 1442–1463.
3. Rao, Y. J. 1997. In-fibre Bragg grating sensors. *Measurement Science and Technology*, 8(4), 355.
4. Hill, K. O., Malo, B., Bilodeau, F., Johnson, D. C., and Albert, J. 1993. Bragg gratings fabricated in monomode photosensitive optical fiber by UV exposure through a phase mask. *Applied Physics Letters*, 62(10), 1035–1037.
5. Meltz, G., Morey, W., and Glenn, W. H. 1989. Formation of Bragg gratings in optical fibers by a transverse holographic method. *Optics Letters*, 14(15), 823–825.
6. Martinez, A., Dubov, M., Khrushchev, I., and Bennion, I. 2004. Direct writing of fibre Bragg gratings by femtosecond laser. *Electronics Letters*, 40(19), 1170–1172.
7. Malo, B., Hill, K. O., Bilodeau, F., Johnson, D. C., and Albert, J. 1993. Point-by-point fabrication of micro-Bragg gratings in photosensitive fibre using single excimer pulse refractive index modification techniques. *Electronics Letters*, 29(18), 1668–1669.
8. Hill, K. O. and Meltz, G. 1997. Fiber Bragg grating technology fundamentals and overview. *Journal of Lightwave Technology*, 15(8), 1263–1276.
9. Erdogan, T. 1997. Fiber grating spectra. *Journal of Lightwave Technology*, 15(8), 1277–1294.
10. Wan, X. and Taylor, H. F. 2002. Intrinsic fiber Fabry–Perot temperature sensor with fiber Bragg grating mirrors. *Optics Letters*, 27(16), 1388–1390.
11. Niu, S., Liao, Y., Yao, Q., and Hu, Y. 2012. Resolution and sensitivity enhancements in strong grating based fiber Fabry–Perot interferometric sensor system utilizing multiple reflection beams. *Optics Communications*, 285(12), 2826–2831.

12. Kaddu, S. C. 1995. Low coherence fibre optic Fabry–Perot sensors suitable for multiplexed strain measurement. PhD thesis, Victoria University, Melbourne.
13. Wang, Z., Shen, F., Song, L., Wang, X., and Wang, A. 2007. Multiplexed fiber Fabry–Perot interferometer sensors based on ultrashort Bragg gratings. *IEEE Photonics Technology Letters*, 19(8), 622–624.
14. Rao, Y. J., Cooper, M. R., Jackson, D. A., Pannell, C. N., and Reekie, L. 2000. Absolute strain measurement using an in-fibre-Bragg-grating-based Fabry–Perot sensor. *Electronics Letters*, 36(8), 708–709.
15. Kaddu, S. C., Booth, D. J., Garchev, D. D., and Collins, S. F. 1997. Intrinsic fibre Fabry–Perot sensors based on co-located Bragg gratings. *Optics Communications*, 142(4), 189–192.
16. Yulianti, I., Supa'at, A. S. M., Idrus, S. M., and Anwar, M. R. S. 2013. Design of fiber Bragg grating-based Fabry–Perot sensor for simultaneous measurement of humidity and temperature. *Optik—International Journal for Light and Electron Optics*, 124(19), 3919–3923.
17. Hill, K. O., Bilodeau, F., Malo, B., Kitagawa, T., Thériault, S., Johnson, J., Albert, J., and Takiguchi, K. 1994. Chirped in-fiber Bragg gratings for compensation of optical-fiber dispersion. *Optics Letters*, 19(17), 1314–1316.
18. Wada, A., Ikuma, K., Syoji, M., Tanaka, S., and Takahashi, N. 2013. Wide-dynamic-range high-resolution fiber Fabry–Perot interferometric sensor with chirped fiber Bragg gratings. *Journal of Lightwave Technology*, 31(19), 3176–3180.
19. Lee, C. E., Atkins, R. A., and Taylor, H. F. 1988. Performance of a fiber-optic temperature sensor from –200 to 1050°C. *Optics Letters*, 13(11), 1038–1040.
20. Choi, H. S., Taylor, H. F., and Lee, C. E. 1997. High-performance fiber-optic temperature sensor using low-coherence interferometry. *Optics Letters*, 22(23), 1814–1816.
21. Tseng, S. M. and Chen, C. L. 1988. Optical fiber Fabry–Perot sensors. *Applied Optics*, 27(3), 547–551.
22. Mathew, J., Schneller, O., Polyzos, D., Havermann, D., Carter, R., MacPherson, W., Hand, D., and Maier, R. 2015. In-fibre Fabry–Perot cavity sensor for high temperature applications. *IEEE/OSA Journal Lightwave Technology*, 33(12), 2419–2425.
23. Ohtsu, M. 1995. Progress of high-resolution photon scanning tunneling microscopy due to a nanometric fiber probe. *Journal of Lightwave Technology*, 3(7), 1200–1221.
24. Mononobe, S., Maheswari, R. U., and Ohtsu, M. 1997. Fabrication of a pencil-shaped fiber probe with a nanometric protrusion from a metal film for near-field optical microscopy. *Optics Express*, 1(8), 229–233.
25. Tso, S. T. and Pask, J. A. 1982. Reaction of glasses with hydrofluoric acid solution. *Journal of the American Ceramic Society*, 65(7), 360–362.
26. Liang, D. T. and Readey, D. W. 1987. Dissolution kinetics of crystalline and amorphous silica in hydrofluoric–hydrochloric acid mixtures. *Journal of the American Ceramic Society*, 70(8), 570–577.

27. Chen, X., Shen, F., Wang, Z., Huang, Z., and Wang, A. 2006. Micro-air-gap based intrinsic Fabry–Perot interferometric fiber-optic sensor. *Applied Optics*, 45(30), 7760–7766.
28. Machavaram, V. R., Badcock, R. A., and Fernando, G. F. 2007. Fabrication of intrinsic fibre Fabry–Perot sensors in silica fibres using hydrofluoric acid etching. *Sensors and Actuators A: Physical*, 138(1), 248–260.
29. Tsai, W. H., and Lin, C. J. 2001. A novel structure for the intrinsic Fabry–Perot fiber-optic temperature sensor. *Journal of Lightwave Technology*, 19(5), 682.
30. Huang, Z., Zhu, Y., Chen, X., and Wang, A. 2005. Intrinsic Fabry–Perot fiber sensor for temperature and strain measurements. *IEEE Photonics Technology Letters*, 17(11), 2403–2405.
31. Tan, X., Geng, Y., Li, X., Gao, R., and Yin, Z. 2013. High temperature microstructured fiber sensor based on a partial-reflection-enabled intrinsic Fabry–Perot interferometer. *Applied Optics*, 52(34), 8195–8198.
32. Wang, W., Hu, Z., Ma, M., Lin, H., and Hu, Y. 2014. Strong fiber Bragg grating based asymmetric Fabry–Perot sensor system with multiple reflections for high sensitivity enhancement. *Optical Fiber Technology*, 20(2), 95–99.
33. Duan, D. W., Rao, Y. J., Wen, W. P., Yao, J., Wu, D., Xu, L. C., and Zhu, T. 2011. In-line all-fibre Fabry–Perot interferometer high temperature sensor formed by large lateral offset splicing. *Electronics Letters*, 47(6), 401–403.
34. Lü, T. 2015. Influence of cavity loss on an extrinsic Fabry–Perot cavity intensity-based pressure sensor. *Review of Scientific Instruments*, 86(9), 095002.
35. Rao, Y. J., Jiang, J., and Zhou, C. X. 2005. Spatial-frequency multiplexed fiber-optic Fizeau strain sensor system with optical amplification. *Sensors and Actuators A: Physical*, 120(2), 354–359.
36. Hunger, D., Steinmetz, T., Colombe, Y., Deutsch, C., Hänsch, T. W., and Reichel, J. 2010. A fiber Fabry–Perot cavity with high finesse. *New Journal of Physics*, 12(6), 065038.
37. Wang, W., Wu, N., Tian, Y., Wang, X., Niezrecki, C., and Chen, J. 2009. Optical pressure/acoustic sensor with precise Fabry–Perot cavity length control using angle polished fiber. *Optics Express*, 17(19), 16613–16618.
38. Murphy, K. A., Gunther, M. F., Vengsarkar, A. M., and Claus, R. O. 1991. Quadrature phase-shifted, extrinsic Fabry–Perot optical fiber sensors. *Optics Letters*, 16(4), 273–275.
39. Chin, K. K., Sun, Y., Feng, G., Georgiou, G. E., Guo, K., Niver, E., Roman, H., and Noe, K. 2007. Fabry–Perot diaphragm fiber-optic sensor. *Applied Optics*, 46(31), 7614–7619.
40. Donlagic, D. and Cibula, E. 2005. All-fiber high-sensitivity pressure sensor with SiO2 diaphragm. *Optics Letters*, 30(16), 2071–2073.
41. Wang, W., Yu, Q., Li, F., Zhou, X., and Jiang, X. 2012. Temperature-insensitive pressure sensor based on all-fused-silica extrinsic Fabry–Pérot optical fiber interferometer. *IEEE Sensors Journal*, 12(7), 2425–2429.

42. Chen, J. H., Zhao, J. R., Huang, X. G., and Huang, Z. J. 2010. Extrinsic fiber-optic Fabry–Perot interferometer sensor for refractive index measurement of optical glass. *Applied Optics*, 49(29), 5592–5596.
43. Chen, J. H., Huang, X. G., Zhao, J. R., Tao, J., He, W. X., and Liu, S. H. 2010. Fabry–Perot interference-based fiber-optic sensor for small displacement measurement. *Optics Communications*, 283(17), 3315–3319.
44. Marchman, H. M., Griffith, J. E., and Filas, R. W. 1994. Fabrication of optical fiber probes for nanometer-scale dimensional metrology. *Review of Scientific Instruments*, 65(8), 2538–2541.
45. Mononobe, S. and Ohtsu, M. 1998. Development of a fiber used for fabricating application oriented near-field optical probes. *IEEE Photonics Technology Letters*, 10(1), 99–101.
46. Tuck, C. J., Hague, R., and Doyle, C. 2006. Low cost optical fibre based Fabry–Perot strain sensor production. *Measurement Science and Technology*, 17(8), 2206.
47. Gong, Y., Rao, Y. J., Guo, Y., Ran, Z. L., and Wu, Y. 2009. Temperature-insensitive micro Fabry–Pérot strain sensor fabricated by chemically etching er-doped fiber. *IEEE Photonics Technology Letters*, 21(22), 1725–1727.
48. Ran, Z. L., Rao, Y. J., Deng, H. Y., and Liao, X. 2007. Miniature in-line photonic crystal fiber etalon fabricated by 157 nm laser micromachining. *Optics Letters*, 32(21), 3071–3073.
49. Ran, Z., Liu, Z., Rao, Y., Xu, F., Sun, D., Yu, X., Xu, B., and Zhang, J. 2011. Miniature fiber-optic tip high pressure sensors micromachined by 157 nm laser. *IEEE Sensors Journal*, 11(5), 1103–1106.
50. Ran, Z., Rao, Y., Zhang, J., Liu, Z., and Xu, B. 2009. A miniature fiber-optic refractive-index sensor based on laser-machined Fabry–Perot interferometer tip. *Journal of Lightwave Technology*, 27(23), 5426–5429.
51. Rao, Y. J., Deng, M., Duan, D. W., Yang, X. C., Zhu, T., and Cheng, G. H. 2007. Micro Fabry–Perot interferometers in silica fibers machined by femtosecond laser. *Optics Express*, 15(21), 14123–14128.
52. Wei, T., Han, Y., Tsai, H. L., and Xiao, H. 2008. Miniaturized fiber inline Fabry–Perot interferometer fabricated with a femtosecond laser. *Optics Letters*, 33(6), 536–538.
53. Ran, Z. L., Rao, Y. J., Liao, X., and Deng, H. Y. 2009. Self-enclosed all-fiber in-line etalon strain sensor micromachined by 157-nm laser pulses. *Journal of Lightwave Technology*, 27(15), 3143–3149.
54. Wei, T., Han, Y., Li, Y., Tsai, H. L., and Xiao, H. 2008. Temperature-insensitive miniaturized fiber inline Fabry–Perot interferometer for highly sensitive refractive index measurement. *Optics Express*, 16(8), 5764–5769.
55. Yuan, W., Wang, F., Savenko, A., Petersen, D. H., and Bang, O. 2011. Note: Optical fiber milled by focused ion beam and its application for Fabry–Pérot refractive index sensor. *Review of Scientific Instruments*, 82(7), 076103.
56. Wang, D. H., Wang, S. J., and Jia, P. G. 2012. In-line silica capillary tube all-silica fiber-optic Fabry–Perot interferometric sensor for detecting high intensity focused ultrasound fields. *Optics Letters*, 37(11), 2046–2048.

57. Cao, Z., Jiang, L., Wang, S., Wang, M., Liu, D., Wang, P., Zhang, F., and Lu, Y. 2015. All-glass extrinsic Fabry–Perot interferometer thermo-optic coefficient sensor based on a capillary bridged two fiber ends. *Applied Optics*, 54(9), 2371–2375.
58. Liu, S., Yang, K., Wang, Y., Qu, J., Liao, C., He, J. et al. 2015. High-sensitivity strain sensor based on in-fiber rectangular air bubble. *Scientific Reports*, 5, 7624.
59. Jiang, Y. and Tang, C. 2008. High-finesse micro-lens fiber-optic extrinsic Fabry–Perot interferometric sensors. *Smart Materials and Structures*, 17(5), 055013.
60. Chen, F., Jiang, Y., Gao, H., and Jiang, L. 2015. A high-finesse fiber optic Fabry–Perot interferometer based magnetic-field sensor. *Optics and Lasers in Engineering*, 71, 62–65.
61. Liu, J., Sun, Y., and Fan, X. 2009. Highly versatile fiber-based optical Fabry–Pérot gas sensor. *Optics Express*, 17(4), 2731–2738.
62. Zhao, J. R., Huang, X. G., He, W. X., and Chen, J. H. 2010. High-resolution and temperature-insensitive fiber optic refractive index sensor based on Fresnel reflection modulated by Fabry–Perot interference. *Journal of Lightwave Technology*, 28(19), 2799–2803.
63. Chen, J. H., Huang, X. G., and Huang, Z. J. 2010. Simple thin-film fiber optic temperature sensor based on Fabry–Perot interference. *Optical Engineering*, 49(4), 044402.
64. Jiang, M. and Gerhard, E. 2001. A simple strain sensor using a thin film as a low-finesse fiber-optic Fabry–Perot interferometer. *Sensors and Actuators A: Physical*, 88(1), 41–46.
65. Wu, X. and Solgaard, O. 2013. Short-cavity multimode fiber-tip Fabry-Pérot sensors. *Optics Express*, 21(12), 14487–14499.
66. Zhao, T., Gong, Y., Rao, Y., Wu, Y., Ran, Z., and Wu, H. 2011. Fiber-optic Fabry–Perot strain sensor based on graded-index multimode fiber. *Chinese Optics Letters*, 9(5), 050602.
67. Beard, P. C. and Mills, T. N. 1996. Extrinsic optical-fiber ultrasound sensor using a thin polymer film as a low-finesse Fabry–Perot interferometer. *Applied Optics*, 35(4), 663–675.
68. Roriz, P., Frazão, O., Lobo-Ribeiro, A. B., Santos, J. L., and Simões, J. A. 2013. Review of fiber-optic pressure sensors for biomedical and biomechanical applications. *Journal of Biomedical Optics*, 18(5), 050903.
69. Ma, C., Dong, B., Lally, E. M., and Wang, A. 2012. Optimization of single-/multi-/single-mode intrinsic Fabry–Perot fiber sensors. *Journal of Lightwave Technology*, 30(14), 2281–2288.
70. Ma, C. and Wang, A. 2010. Multimode excitation-induced phase shifts in intrinsic Fabry–Perot interferometric fiber sensor spectra. *Applied Optics*, 49(25), 4836–4845.
71. Ran, Z. L., Rao, Y. J., Liu, W. J., Liao, X., and Chiang, K. S. 2008. Laser-micromachined Fabry–Perot optical fiber tip sensor for high-resolution temperature-independent measurement of refractive index. *Optics Express*, 16(3), 2252–2263.

72. Choi, H. Y., Park, K. S., Park, S. J., Paek, U. C., Lee, B. H., and Choi, E. S. 2008. Miniature fiber-optic high temperature sensor based on a hybrid structured Fabry–Perot interferometer. *Optics Letters*, 33(21), 2455–2457.
73. Wang, R. and Qiao, X. 2014. Hybrid optical fiber Fabry–Perot interferometer for simultaneous measurement of gas refractive index and temperature. *Applied Optics*, 53(32), 7724–7728.
74. Gong, Y., Zhao, T., Rao, Y. J., Wu, Y., and Guo, Y. 2010. A ray-transfer-matrix model for hybrid fiber Fabry–Perot sensor based on graded-index multimode fiber. *Optics Express*, 18(15), 15844–15852.
75. Gong, Y., Guo, Y., Rao, Y.J., Zhao, T., Wu, Y., Ran, Z. L. 2011. Sensitivity analysis of hybrid fiber Fabry–Pérot refractive-index sensor. *Acta Physica Sinica*, 60(6), 064202.
76. Gong, Y., Guo, Y., Rao, Y. J., Zhao, T., and Wu, Y. 2010. Fiber-optic Fabry–Perot sensor based on periodic focusing effect of graded-index multimode fibers. *IEEE Photonics Technology Letters*, 22(23), 1708–1710.
77. Rao, Y. J., Ran, Z. L., Liao, X., and Deng, H. Y. 2007. Hybrid LPFG/MEFPI sensor for simultaneous measurement of high-temperature and strain. *Optics Express*, 15(22), 14936–14941.
78. Frazao, O., Silva, S. F., Viegas, J., Baptista, J. M., Santos, J. L., and Roy, P. 2010. A hybrid Fabry–Perot/Michelson interferometer sensor using a dual asymmetric core microstructured fiber. *Measurement Science and Technology*, 21(2), 025205.
79. Knight, J. C., Birks, T. A., Russell, P. S. J., and Atkin, D. M. 1996. All-silica single-mode optical fiber with photonic crystal cladding. *Optics Letters*, 21(19), 1547–1549.
80. Birks, T. A., Knight, J. C., and Russell, P. S. J. 1997. Endlessly single-mode photonic crystal fiber. *Optics Letters*, 22(13), 961–963.
81. Zhu, T., Ke, T., Rao, Y., and Chiang, K. S. 2010. Fabry–Perot optical fiber tip sensor for high temperature measurement. *Optics Communications*, 283(19), 3683–3685.
82. Zhang, J., Sun, H., Rong, Q., Ma, Y., Liang, L., Xu, Q., Zhao, P., Feng, Z., Hu, M., and Qiao, X. 2012. High-temperature sensor using a Fabry–Perot interferometer based on solid-core photonic crystal fiber. *Chinese Optics Letters*, 10(7), 070607.
83. Rao, Y. J., Zhu, T., Yang, X. C., and Duan, D. W. 2007. In-line fiber-optic etalon formed by hollow-core photonic crystal fiber. *Optics Letters*, 32(18), 2662–2664.
84. Ferreira, M. S., Bierlich, J., Kobelke, J., Schuster, K., Santos, J. L., and Frazão, O. 2012. Towards the control of highly sensitive Fabry–Pérot strain sensor based on hollow-core ring photonic crystal fiber. *Optics Express*, 20(20), 21946–21952.
85. Shi, Q., Lv, F., Wang, Z., Jin, L., Hu, J. J., Liu, Z., Kai, G., and Dong, X. 2008. Environmentally stable Fabry–Pérot-type strain sensor based on hollow-core photonic bandgap fiber. *IEEE Photonics Technology Letters*, 20(4), 237–239.

86. Jin, L., Guan, B. O., and Wei, H. 2013. Sensitivity characteristics of Fabry–Perot pressure sensors based on hollow-core microstructured fibers. *Journal of Lightwave Technology*, 31(15), 2526–2532.
87. Rao, Y. J., Deng, M., Duan, D. W., and Zhu, T. 2008. In-line fiber Fabry–Perot refractive-index tip sensor based on endlessly photonic crystal fiber. *Sensors and Actuators A: Physical*, 148(1), 33–38.
88. Jiang, M. S., Sui, Q. M., Jin, Z. W., Zhang, F. Y., and Jia, L. 2014. Temperature-independent optical fiber Fabry–Perot refractive-index sensor based on hollow-core photonic crystal fiber. *Optik—International Journal for Light and Electron Optics*, 125(13), 3295–3298.
89. Wu, D., Huang, Y., Fu, J. Y., and Wang, G. Y. 2015. Fiber Fabry–Perot tip sensor based on multimode photonic crystal fiber. *Optics Communications*, 338, 288–291.
90. Wu, D., Huang, W., Wang, G. Y., Fu, J. Y., and Chen, Y. Y. 2014. In-line fiber Fabry–Perot refractive index tip sensor based on photonic crystal fiber and spectrum differential integration method. *Optics Communications*, 313, 270–275.
91. Wu, D. K., Kuhlmey, B. T., and Eggleton, B. J. 2009. Ultrasensitive photonic crystal fiber refractive index sensor. *Optics Letters*, 34(3), 322–324.
92. Deng, M., Tang, C. P., Zhu, T., Rao, Y. J., Xu, L. C., and Han, M. 2010. Refractive index measurement using photonic crystal fiber-based Fabry–Perot interferometer. *Applied Optics*, 49(9), 1593–1598.
93. Frazão, O., Aref, S. H., Baptista, J. M., Santos, J. L., Latifi, H., Farahi, F., Kobelke, J., and Schuster, K. 2009. Fabry–Pérot cavity based on a suspended-core fiber for strain and temperature measurement. *IEEE Photonics Technology Letters*, 21(17), 1229–1231.
94. Pinto, A. M. R., Frazão, O., Santos, J. L., Lopez-Amo, M., Kobelke, J., and Schuster, K. 2010. Interrogation of a suspended-core Fabry–Perot temperature sensor through a dual wavelength Raman fiber laser. *IEEE Photonics Technology Letters*, 28(21), 3149–3155.
95. Frazao, O., Baptista, J. M., Santos, J. L., Kobelke, J., and Schuster, K. 2009. Refractive index tip sensor based on Fabry–Perot cavities formed by a suspended core fibre. *Journal of the European Optical Society—Rapid Publications*, 4(4), 09041.
96. Uchimura, R., Wada, A., Tanaka, S., and Naoyuki, T. 2015 Fiber Fabry–Perot interferometric sensor using Bragg Gratings in polarization maintaining fiber. *Journal of Lightwave Technology*, 33(12), 2499–2503.
97. Tian, J., Lu, Y., Zhang, Q., and Han, M. 2013. Microfluidic refractive index sensor based on an all-silica in-line Fabry–Perot interferometer fabricated with microstructured fibers. *Optics Express*, 21(5), 6633–6639.
98. Tian, J., Zhang, Q., Fink, T., Li, H., Peng, W., and Han, M. 2012. Tuning operating point of extrinsic Fabry–Perot interferometric fiber-optic sensors using microstructured fiber and gas pressure. *Optics Letters*, 37(22), 4672–4674.

99. Pevec, S. and Donlagic, D. 2011. All-fiber, long-active-length Fabry–Perot strain sensor. *Optics Express*, 19(16), 15641–15651.
100. Gong, Y., Yu, C. B., Wang, T. T., Liu, X. P., Wu, Y., Rao, Y. J., Zhang, M. L., Wu, H. J., Cheng, X. X., and Peng, G. D. 2014. Highly sensitive force sensor based on optical microfiber asymmetrical Fabry–Perot interferometer. *Optics Express*, 22(3), 3578–3584.
101. Zhang, J., Sun, Q., Liang, R., Wo, J., Liu, D., and Shum, P. 2012. Microfiber Fabry–Perot interferometer fabricated by taper-drawing technique and its application as a radio frequency interrogated refractive index sensor. *Optics Letters*, 37(14), 2925–2927.
102. Li, J., Shen, X., Sun, L. P., and Guan, B. O. 2013. Characteristics of microfiber Fabry–Perot resonators fabricated by UV exposure. *Optics Express*, 21(10), 12111–12121.
103. Kou, J. L., Feng, J., Ye, L., Xu, F., and Lu, Y. Q. 2010. Miniaturized fiber taper reflective interferometer for high temperature measurement. *Optics Express*, 18(13), 14245–14250.

3
Fabrication Techniques for FFPI Sensors

FFPI sensors can be formed based on a variety of structures [1–5]. In this chapter, various technologies are described for the fabrication of FFPI sensors. Generally, the FP cavities can be formed by manually splicing a hollow-core fiber section between SMF and a diaphragm; by splicing photonic crystal fiber (PCF); and by splicing a section of sapphire fiber to SMF. Manual fabrication of FFP sensors is time consuming and labor intensive, and prone to contamination and damage [5]. In recent years, the new demand for novel sensors such as high-temperature strain and pressure sensors and miniature sensors used in tiny spaces is growing continuously. Novel fabrication techniques for fiber-optic sensors are very much in demand to create new functional sensors. Generally, in-line miniature sensors fabricated directly on optical fiber are difficult to be mass produced with good reproducibility due to the miniature size and extreme hardness of the optical fiber. Femtosecond (fs) lasers, focused ion beam (FIB), 157-nm excimer lasers, and chemical etching of doped fibers are good candidates for fabricating microsensors on fibers.

3.1 Femtosecond Laser Micromachining

3.1.1 Open Notch FFP Sensor Fabricated by Femtosecond Laser

The advancement in fs-laser technology has opened a new window of opportunity for precise fabrication of microdevices with true three-dimensional (3D) configurations. Fs lasers could emit ultra-intense laser pulses, offering the ability to ablate most solid materials with high surface quality, and/or to change their refractive index without special pretreating and doping. Fs laser pulses with extremely high peak power produce almost no thermal damage as the pulse duration is shorter than the thermalization time [6]. Due to the multiphoton nature of the interaction, the ablation process can be conducted on the material surface as well as within its bulk. Fs lasers have been successfully used for directly writing optical waveguides [7,8] and micromachining microchannels and microchambers in glasses [9].

When a fs laser is focused onto optical fiber, the energy absorption takes place through nonlinear phenomena such as multiphoton absorption (a number of photons could be absorbed simultaneously, when material is irradiated by high-power laser light beam), tunneling, and avalanche ionization. As long as the absorbed energy is high enough, catastrophic material damage occurs, which leads to the formation of special structures [10]. Intensive index modulation of up to 0.001 could be induced in fiber by fs laser. If the absorbed energy increases further, a microstructure in optical fiber can be formed with designed shape and size.

Fiber-optic micro extrinsic FP interferometers (MEFPIs) could be fabricated by using fs laser to mill the fiber. The schematic diagram of fs-laser micromachining system is shown in Figure 3.1. A home-built chirped-pulse amplified Ti:sapphire laser system with high pulse energy of up to 100 μJ for 100 fs pulses with a repetition rate of 1–5 kHz was used. After being reflected from a dichroic mirror, the light was

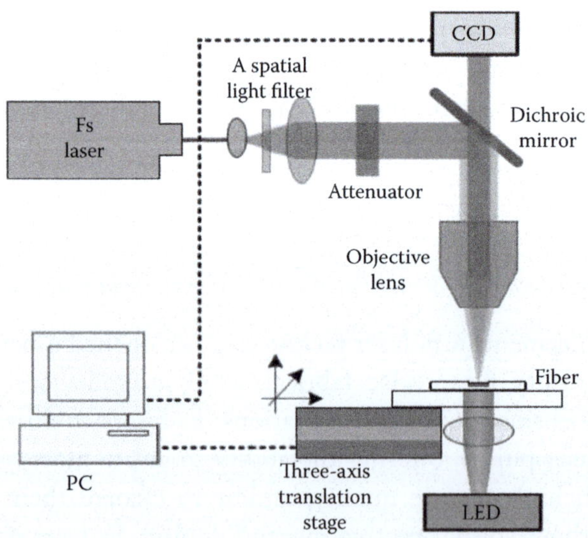

Figure 3.1 Schematic diagram of the fs-laser system. (From Rao, Y. J. et al. 2007. *Optics Express*, 15(21), 14123–14128.)

focused onto the silica fiber by an objective lens (50× magnification, 0.65 numerical aperture [NA]). A light-emitting diode (LED) was used to illuminate the sample so that the MFPI sensor after ablation can be monitored in real time by using a charge-coupled device (CCD) camera attached to a phase-contrast optical microscope. A computer-controlled three-axis translation stage (100 nm resolution at X direction, 125 nm at Y direction, and 7 nm at Z direction; PI, German) was employed to carry out the required movements of the silica fiber.

The laser wavelength was 800 nm with a pulse width of 120 fs and a repetition rate of 1 kHz. The focused spot size was 5 μm and the pulse energy was 20 μJ. The SMF (Corning: SMF-28) and the PCF (Crystal Fiber: ESM-12-01) were mounted, respectively, onto the translation stage and moved at a speed of ~300 μm/s. A single-pass exposure over an area of 80 μm × 30 μm was carried out and such a process was repeated by several times until the ablated FP cavity formed to meet the requirements [11].

Figure 3.2 displays the optical micrograph of an MFPI with an 80-μm cavity length based on SMF, and the related reflective spectrum is shown in Figure 3.3. It can be seen from Figure 3.3 that the interferometric fringes are good enough for sensing applications, but the fringe visibility is relative low. The poor fringe visibility was mainly caused by three reasons: (1) the light scattering loss at the laser-ablated surface; (2) the non-perpendicular surface orientation with respect to the fiber axis; (3) the coupling loss as a result of recoupling the light reflected from the second end face of the FP cavity back into the fiber core [11].

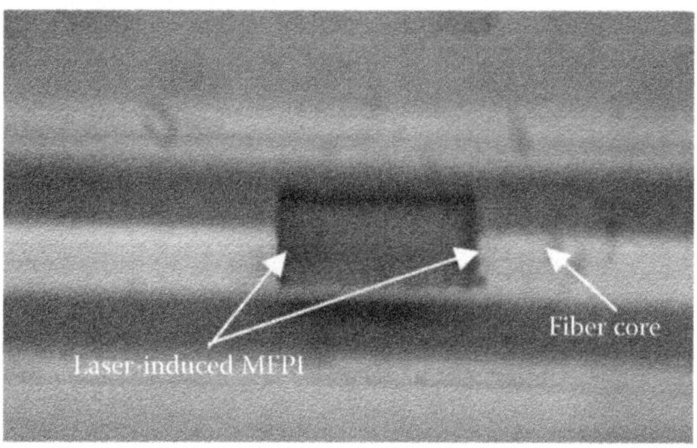

Figure 3.2 Optical micrograph of an MFPI with an 80-μm cavity length on the SMF. (From Rao, Y. J. et al. 2007. *Optics Express*, 15(21), 14123–14128.)

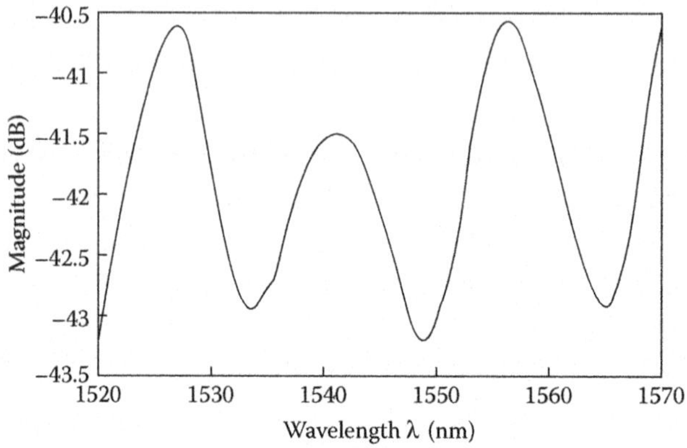

Figure 3.3 Reflective spectrum of an MFPI with an 80-μm cavity length on the SMF. (From Rao, Y. J. et al. 2007. *Optics Express*, 15(21), 14123–14128.)

Figure 3.4 shows the optical micrograph of an MFPI sensor with a 75-μm cavity length based on the PCF. It can be seen that the two sides of such a PCF-based MFPI are neat and parallel. Comparing with the SMF-based MFPI, the fringe visibility of the PCF-based MFPI was improved by a few dB as shown in Figure 3.5; this is mainly due to the following reasons: (i) the PCF is entirely made from pure fused silica, which reduces the sputtered remains as well as light scattering [12]; (ii) the cladding of the PCF is a 2D photonic crystal structure with air holes distributed along the length of the fiber, diffusing the laser-ablation-generated heat and pressure quickly, and hence can effectively reduce the thermal damage on the cross section of the PCF ablated.

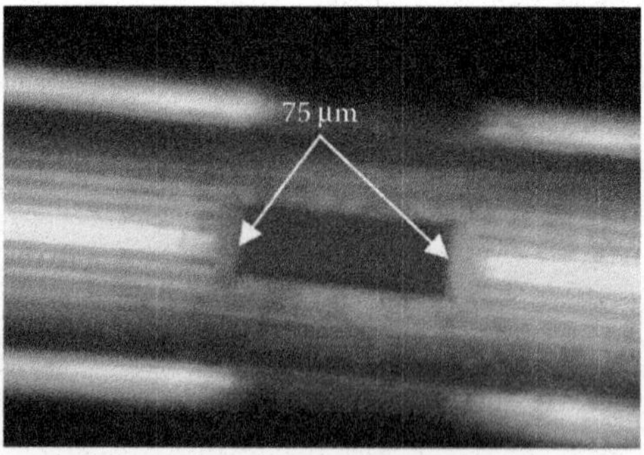

Figure 3.4 Optical micrograph of an MFPI with a 75-μm cavity length based on the PCF. (From Rao, Y. J. et al. 2007. *Optics Express*, 15(21), 14123–14128.)

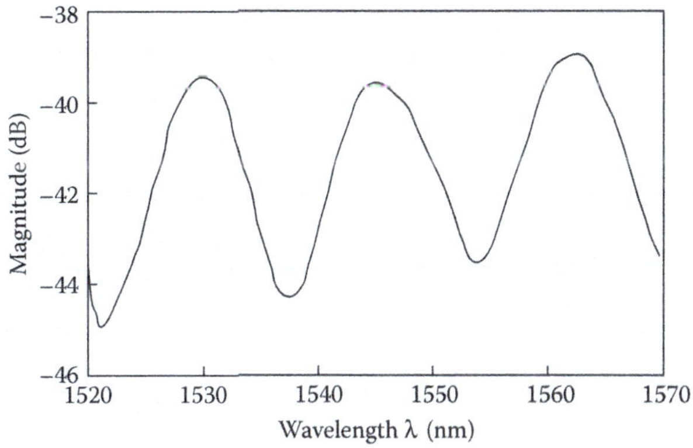

Figure 3.5 Reflective spectrum of the PCF MFPI. (From Rao, Y. J. et al. 2007. *Optics Express*, 15(21), 14123–14128.)

For achieving better fringe visibility, it is believed that the surface roughness can be reduced by reducing the laser scanning steps, of course, at the expense of a long device fabrication time. The non-perpendicular surface orientation can also be minimized by careful adjustment of the stages. The coupling loss increases with the length of the FP cavity. As a result, it may eventually limit the practical length of the cavity. As shown in Figure 3.6, when the cavity length is much shorter (~30 μm), fringe visibility is obviously improved to ~14 dB [13].

The strain responses of the MFPIs based on the SMF in Figure 3.2 and PCF in Figure 3.4 are studied, respectively. The MFPI was fixed on a translation stage with a resolution of 1 μm. The output of the

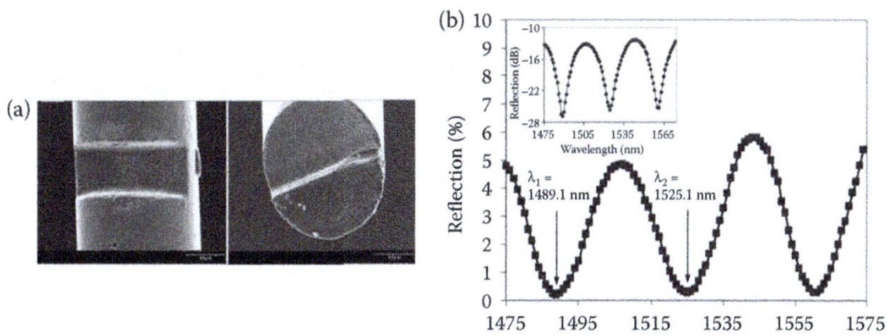

Figure 3.6 (a) SEM images of fiber in-line MFPI device with a ~30-μm cavity length fabricated by fs-laser ablation. (b) Interference spectrum of the EFPI device. (From Wei, T. et al. 2008. *Optics Letters*, 33(6), 536–538.)

tunable laser in the optical spectrum analyzer (OSA) was coupled into the MFPI through a 2 × 2 coupler, and the reflected light was returned via the same coupler to the receiver of the OSA. The tunable laser light is scanned and coupled into the optical device under test. At the same time, a small portion of the laser light is divided by a coupler, passed by a wavelength reference, and finally detected by a photon detector for calibration of the certain wavelength of the tunable laser. When comparing the detected signal from sensor with the reference light, the sensor spectrum could be obtained. The wavelength shift of the SMF-based MFPI sensor and that of the PCF-based MFPI sensor were measured at 1550-nm wavelength band, and their relationships between the wavelength shift and strain variation are shown in Figure 3.7a and b, respectively. Experimental results show that the wavelength–strain sensitivity of the SMF–MFPI sensor and the PCF–MFPI sensor are 0.006 and 0.0045 nm/μm, respectively.

Furthermore, such an open EMFPI could be used to measure refractive index. To evaluate its capability for refractive index measurement, the fiber FPI device was tested using various liquids including methanol, acetone, and isopropanol at room temperature. The interference spectra of the device immersed in various liquids are also shown in Figure 3.8 for comparison. The signal intensity dropped when the device was immersed in liquids as a result of the reduced refractive index contrast and thus lower Fresnel reflections from the cavity end faces. However, the interference fringes maintained a similar visibility. The spectral distance between the two adjacent valleys also decreased, indicating the increase in refractive index of the medium inside the cavity [14].

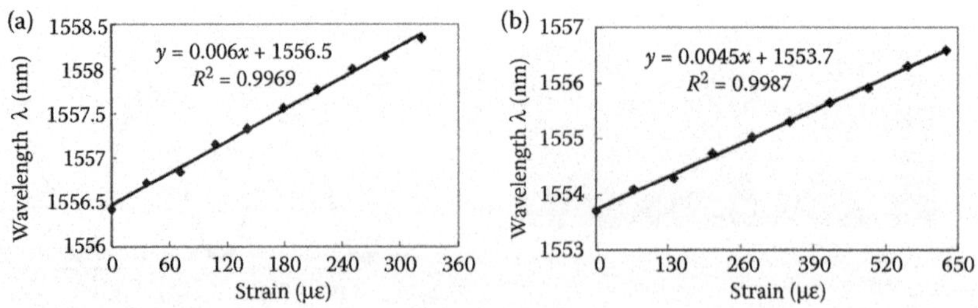

Figure 3.7 (a) Wavelength–strain relationship of the SMF–MFPI sensor. (b) Wavelength–strain relationship of the PCF–MFPI sensor. (From Rao, Y. J. et al. 2007. *Optics Express*, 15(21), 14123–14128.)

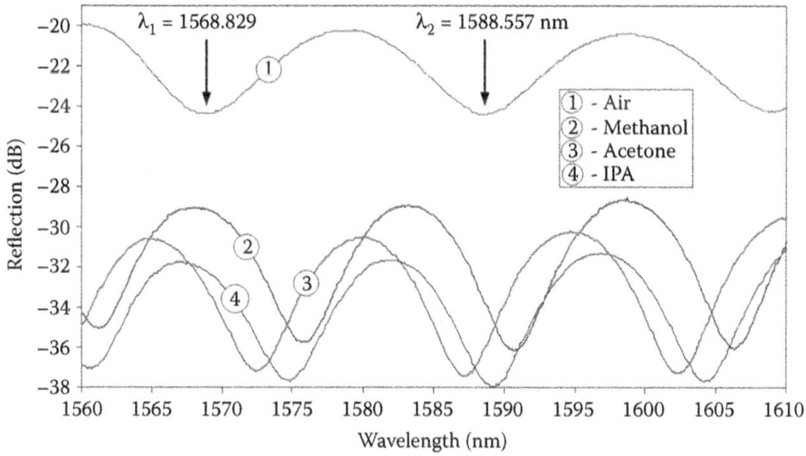

Figure 3.8 Interference spectra of the FPI device in air, methanol, acetone, and isopropanol. (From Wei, T. et al. 2008. *Optics Express*, 16(8), 5764–5769.)

3.1.2 Sealed Fiber-Optic EFPI Fabricated by Femtosecond Laser

A sealed fiber-optic EFPI was also fabricated by using fs-laser micromachining [15,16]. In the device fabrication, fs-laser pulses ($\lambda = 800$ nm) of 120 fs at the repetition rate of 1 kHz were focused onto the fiber end facet by a 20× objective lens with an NA value of 0.5 and a working distance of 2.1 mm. The pulse energy used in the experiment was ~2 μJ. A CCD camera was employed to monitor the fabrication process and record the sample morphology. A section of standard SMF-28 with the core diameter of 8.2 μm and the nominal effective RI of 1.4682 (at 1550 nm) was mounted on a computer-controlled 3D translation stage with a 40-nm resolution. The following steps were adopted in the fabrication process, as illustrated in Figure 3.9. First, fs laser drilled a micro-hole of ~1 μm in diameter at the center of the cleaved fiber end facet, and the fiber tip obtained was then spliced together with another cleaved SMF tip without micro-hole by a fusion splicer (ERICSSON FSU975), with a fusing current and a fusing duration of 16.3 mA and 2.0 s, respectively. The hollow sphere formed had a diameter of ~60 μm. Such a sealed EFPI could be used as strain sensor like the open EFPI.

Based on the sealed EFPI, a refractive index sensor could be formed by drilling a microchannel vertically crossing the FP cavity, which allowed the RI liquid to readily flow in or out of the cavity,

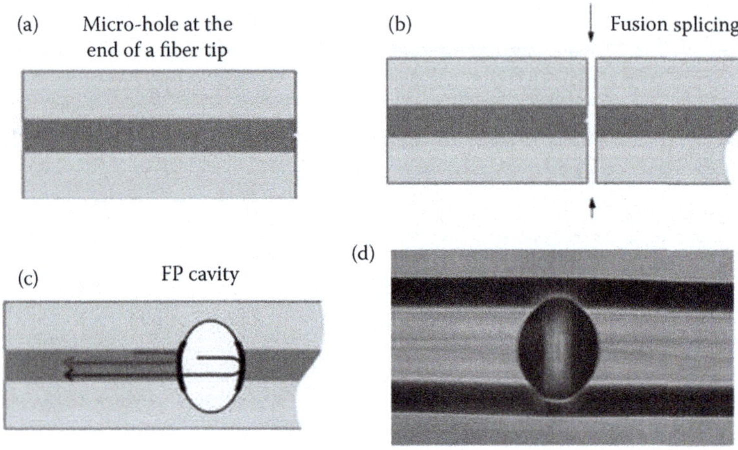

Figure 3.9 Fiber in-line FPI cavity fabrication process. (a) The fs laser creates a micro-hole of ~1 μm in diameter at the center of cleaved fiber end facet. (b) The fiber tip with the micro-hole spliced together with another cleaved SMF tip. (c) FP cavity formed. (d) Microscope image of the fiber in-line FPI cavity. (From Liao, C. R. et al. 2012. *Optics Express*, 20(20), 22813–22818.)

as given in Figure 3.10. In the microchannel fabrication process, the laser beam was firstly focused on the top fiber surface at the microcavity position, scanned at a speed of 2 μm/s with a scanning distance of 40 μm, in parallel to the fiber core axis. After one scanning cycle, the laser beam was shifted by 10 μm, in perpendicular to the fiber axis until a 40 μm × 40 μm square area was drilled through, to create the top part of the microchannel. The fiber was then rotated by 180° to allow fabrication of the bottom part of the microcavity, following the

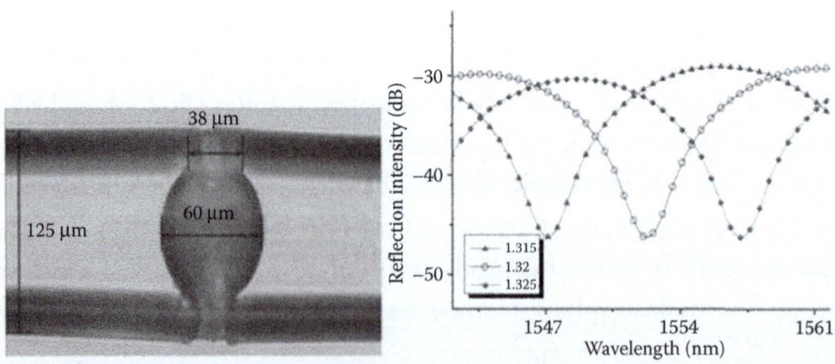

Figure 3.10 Microscope image of the fiber in-line FPI cavity with the microchannel (left). The reflection spectrum at RI = 1.315, 1.32, and 1.325, respectively. (From Liao, C. R. et al. 2012. *Optics Express*, 20(20), 22813–22818.)

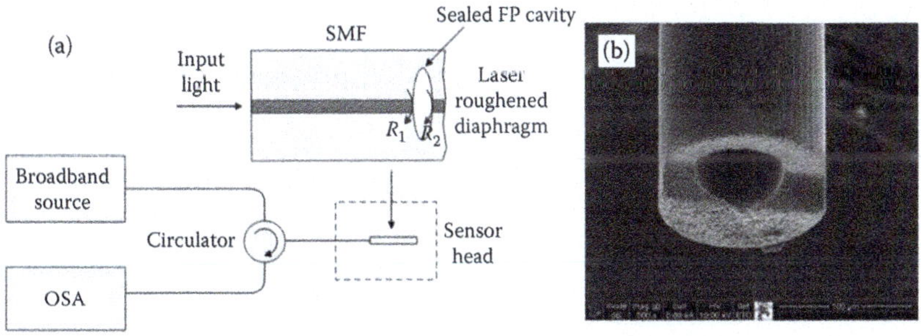

Figure 3.11 (a) Schematic of the pressure sensor. (b) SEM image of the cut-out view of the sensor head. (From Zhang, Y. et al. 2013. *Optics Letters*, 38(22), 4609–4612.)

same procedure as mentioned above. The RI sensitivity obtained is ~994 nm/RIU.

A pressure sensor based on the sealed EFPI was also demonstrated. Figure 3.11 schematically illustrates the structure of the sensor. To fabricate the sensor, a micro-hole was first drilled into the cleaved end face of a SMF by fs-laser micromachining. The inner surface of the laser-drilled micro-hole had quasi-distributed structures of submicron sizes and a low optical reflectivity. The hole-drilled SMF was then spliced to another SMF to form a sealed air cavity in between two fibers. Fusion splicer settings (arc duration, arc power, etc.) were adjusted to avoid the collapsing of the hole. During fusion splice, the rough structure of the micro-hole surface was melted by the arc. As a result, the surface became smooth, and the reflectivity increased. Precision fiber cleaving was applied to cut the fiber so that a thin piece of fiber was left to perform as the diaphragm. Finally, the as-cleaved diaphragm was thinned and roughened by fs laser.

During fs-laser micromachining, the sensor was mounted on a computer-controlled high-precision 3D stage with a resolution of 0.1 μm. Light pulses generated by a regeneratively amplified Ti:sapphire laser (Coherent RegA 9000; 200 fs pulse duration, 250 kHz repetition rate, and 800 nm central wavelength) were focused onto the fiber end face or diaphragm surface through a microscope objective (20× magnification, 0.4 NA). The laser power could be changed by adjusting the laser-beam optics, including a half-wave plate, a polarizer, and several neutral density filters. The actual laser energy used for fabrication was approximately 0.4 μJ per pulse.

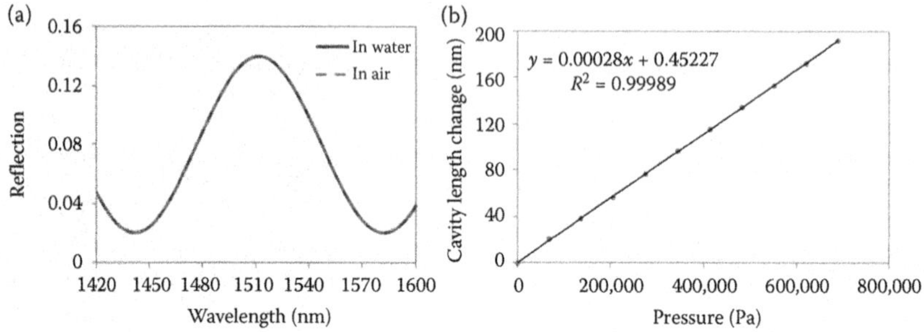

Figure 3.12 (a) Typical interference spectra of the fiber FPI sensor. (b) Pressure-induced interferogram shift of FPI sensor. (From Zhang, Y. et al. 2013. *Optics Letters*, 38(22), 4609–4612.)

The diameter and depth of the hole could be flexibly varied. The diaphragm thinning process was also performed layer-by-layer with a step size of 0.5 μm. The fabrication was completed when a preset depth scan was reached. Figure 3.11b shows the scanning electron microscopic (SEM) image of a sensor head. For easy visualization, half of the sensor head was cut out using fs laser to expose the sealed hole and the diaphragm. This particular sensor had a cavity with the diameter of 70 μm and the length of 100 μm.

Figure 3.12a plots the interference fringes of a typical sensor before and after fs-laser ablation. The sensor had a cavity length of about 8 μm and a hole diameter of 75 μm. The cavity length was calculated to be 7.97 μm. Figure 3.12b shows the measurement results obtained from the sensor, where the changes in cavity length are plotted as a function of the applied pressure. Within the pressure range of 6.895×10^5 Pa, the center of the diaphragm deflected linearly as the pressure changed. Based on the linear fitting curve, the sensitivity of the sensor was calculated to be 2.8×10^{-4} nm/Pa.

3.1.3 Intrinsic Fiber-Optic FPI Fabricated by Femtosecond Laser

Above-mentioned fiber-optic EFPI is formed by fs-laser ablation. When fs-laser power is reduced to some extent, fiber-optic intrinsic FPI could be formed by writing weak reflective mirrors using fs laser to induce permanent refractive index change in fiber, as shown in Figure 3.13a. The typical reflectivity of such a structure was about −45 dB and the loss was about 0.02 dB measured by an optical power

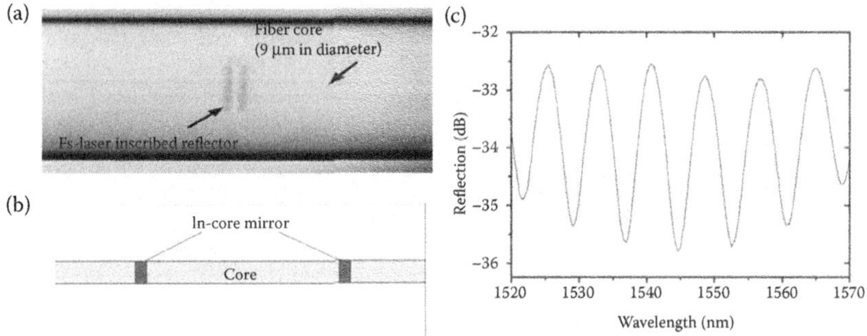

Figure 3.13 (a) Confocal microscopic image of a weak reflector fabricated by fs-laser micromachining. (b) Schematic of the fiber-optic IFPI. (c) The measured reflection spectrum of the fabricated fiber-optic IFPI sensor. (Huang, J. et al. 2014. *Optics Express*, 22(15), 18757–18769; Wang, W. et al. 2010. *Asia Communications and Photonics Conference and Exhibition* (pp. 79900P–79900P). International Society for Optics and Photonics.)

meter. Two cascaded mirrors were written in the core of the fiber; an intrinsic FP interferometer (IFPI) could be achieved, as shown in Figure 3.13b. An IFPI with a cavity length of 100 μm was demonstrated in Figure 3.13c, which could be used as temperature and strain sensors.

3.2 Deep UV Laser Micromachining System

The F_2 laser produces 7.9-eV photons that are known to damage UV-grade fused-silica glasses in long exposures, possibly through absorption involving three- and four-group silicon-ring structures. In low-GeO_2 (i.e., 5%) glasses, the 157-nm photons directly bridge the ~7.1-eV bandgap, which allows access to strong single-photon photosensitivity mechanisms without the need of traditional enhancement techniques.

Using a custom-designed 157-nm excimer laser micromachining system, a series of novel MFFPI sensors are created. Because these micro fiber-optic sensors are directly engraved on silica or sapphire fibers through formation of high-quality, in-line FP etalons, they can offer a number of outstanding advantages, such as very small size, self-temperature compensation, high-temperature (>300°C) survivability, and capability for mass production. These sensors can easily measure strain, pressure, refractive index, temperature, and acceleration in harsh environments.

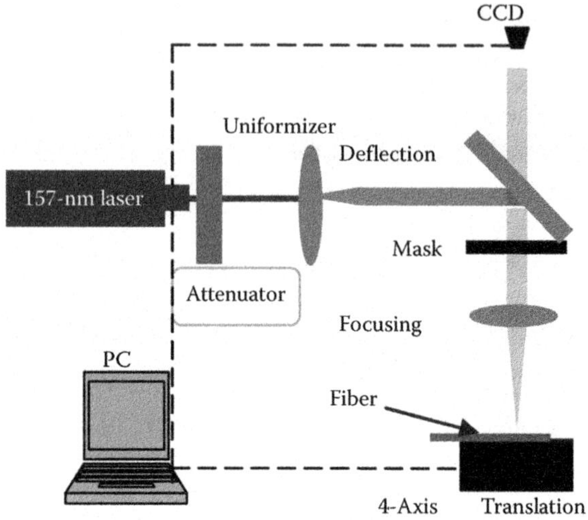

Figure 3.14 Schematic of micromachining system based on 157-nm excimer laser. (From Ran, Z. L. et al. 2007. *Optics Letters*, 32(21), 3071–3073.)

The custom-designed 157-nm laser micromachining system is shown in Figure 3.14. The optical absorption coefficient at 157 nm has a high value of up to ~20,000 cm^{-1}, making it possible to achieve high-quality cool machining of silica or sapphire fibers. The 157-nm laser micromachining system consists of a 157-nm pulsed laser, an optical focusing system with 25× demagnification, and a precise translation stage used to mount the fiber to be engraved. The maximum single pulse energy, pulse width, and pulse repetition rate of the 157-nm laser are 35 mJ, 15 ns, and 50 Hz, respectively.

3.2.1 Direct Engraving of FFPI Strain Sensor on Silica Fibers

Using the 157-nm micromachining system, the first type of MFFPI sensor is fabricated with a micro-rectangular notch structure inside a silica fiber with a typical size of 40–60 lm. The 157-nm source light is projected onto the fiber through a rectangular steel mask that is 25 times larger than the notch. The notch fabricated inside the fiber is an FPI cavity with two smooth and parallel reflective interfaces. An excellent fringe visibility of up to 26 dB was achieved, and a precise strain measurement under high temperatures of up to 800°C was demonstrated (see Figure 3.15). The sensor determines strain by

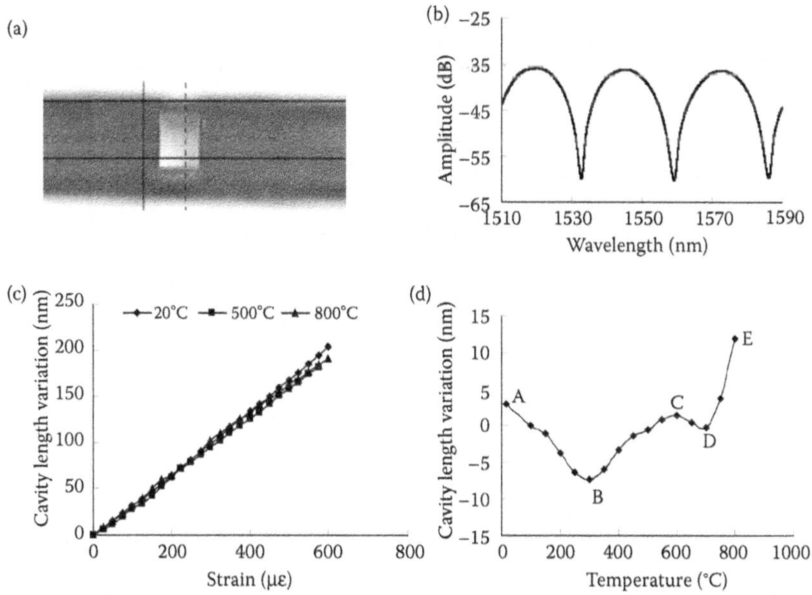

Figure 3.15 (a) SEM photos of the FP sensor; (b) reflective spectrum of PCF FP sensor; (c) strain characteristics of PCF sensor at different temperatures; and (d) temperature response of the sensor. (From Ran, Z. L. et al. 2007. *Optics Letters*, 32(21), 3071–3073.)

measuring the cavity length variation ($\varepsilon = \Delta L/L$, where L and ΔL are the real cavity length and variation to original length, respectively). More importantly, this sensor has the unique feature of temperature self-compensation due to the unique cavity structure. The thermal expansion of the fiber core reduces the cavity length since the fiber core is free along the longitudinal direction due to the air gap, while the thermal expansion of the fiber cladding increases the cavity length due to the existence of the cladding that remains after laser ablation. The two effects cancel each other when temperature varies.

3.2.2 Self-Enclosed FFPI Strain, Pressure, Refractive Index, and Temperature Sensors

The second type of MFFPI sensor is based on a self-enclosed cavity structure, enabling fabrication of sensors that can measure high temperatures, strain, pressure, and refractive index. To form the self-enclosed MFFPI cavity, a circular micro-hole with a typical depth and diameter of tens of micrometers is first made at the tip of a cleaved optical fiber via laser micromachining (see Figure 3.16). Next, an in-line MFFPI cavity is formed and enclosed by simply splicing the engraved fiber to another cleaved fiber, so the cavity is automatically enclosed by the cladding of the fiber. This micro FP etalon can be operated at temperatures of up to 800°C due to its robust, stable, and reliable structure. This device is also insensitive to temperature change due to the hollow-core structure of the etalon and the ultralow thermal expansion coefficient of the silica.

The self-enclosed FP sensor, shown in Figure 3.17a, could be directly used as a high-temperature strain sensor. As shown in Figure 3.18, such a strain sensor can also be operated under very high temperature with good linearity.

To add pressure, temperature, and refractive-index measurement capability to this sensor, the fiber near the air-filled FP cavity is cleaved precisely to form a diaphragm, as shown in Figure 3.17b. For pressure measurement, the diaphragm is about several micrometers thick. When pressure is applied to the sensor, the diaphragm is pushed, which shortens the cavity length of the air cavity. Because the diaphragm is flat and of almost uniform thickness, its center deflection ΔL, under applied pressure ΔP, can be calculated by $\Delta L = 3(1 - \mu^2)R^4\Delta P/16Eh^3$, where L, R, and h are the cavity length, the radius, and the thickness of the diaphragm, respectively, E is Young's modulus, and μ is Poisson's ratio. An example is shown in Figure 3.19 where the sensor is used to measure high pressure of up to 60 MPa, and the pressure response shows good linearity (Figure 3.19b).

If the fiber is cleaved near the air cavity with a much longer distance (more than 100 μm) to form an IFPI, as shown in Figure 3.17c, refractive index and temperature monitoring can be obtained simultaneously. There are two cavities: an air cavity and a much longer all-fiber IFPI. Based on this sensor head, refractive index and temperature can be measured by testing the fringe contrast and optical path difference (OPD) of the IFPI, respectively, as shown in Figure 3.16. RI sensitivity of ~44.9 dB/RIU and temperature OPD coefficient of ~2.67 nm/°C are achieved as shown in Figure 3.20.

Finally, the cleaved fiber can be machined once again on its end face by using a designed mask pattern to form an acceleration sensor

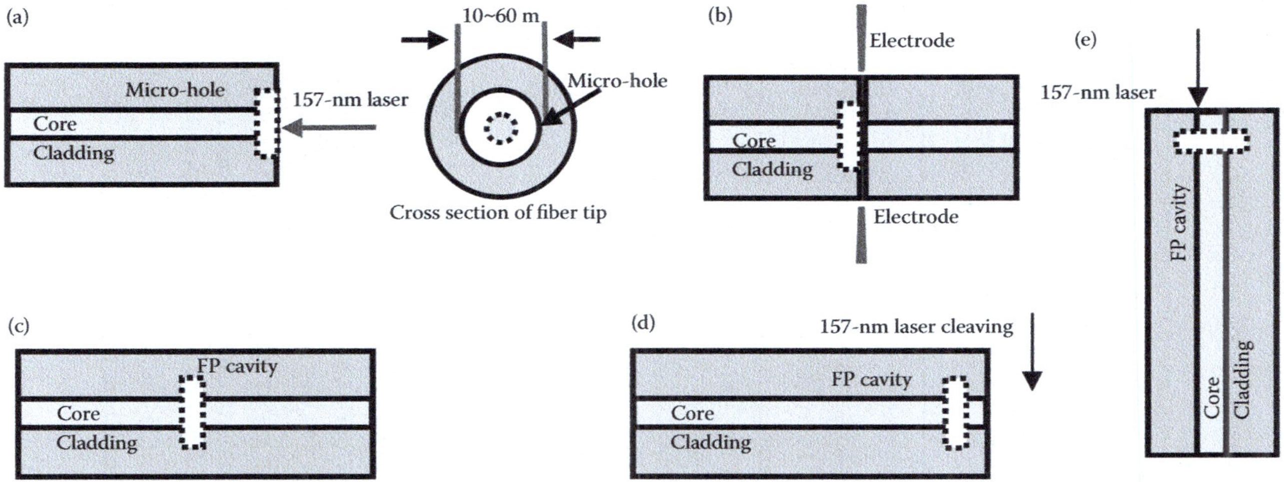

Figure 3.16 Fabrication process of the self-enclosed FFP in-line etalon: (a) creating a circular micro-hole at the fiber tip using a 157-nm laser; (b) splicing the fiber with the micro-hole to another cleaved fiber; (c) completing the FP etalon (strain sensor) after an arc-fusion splicing operation; (d) cleaving the fiber near air FP cavity to form pressure, temperature, and refractive sensors; and (e) fabricating accelerometer sensor. (Rao, Y. J. and Ran, Z. L. 2011. *Laser Focus World*, 47(11), 71; Rao, Y. J. et al. 2007. *Optics Express*, 15(22), 14936–14941; Ran, Z. L. et al. 2008. *Optics Express*, 16(3), 2252–2263; Ran, Z. et al. 2013. *IEEE Sensors Journal*, 13(5), 1988–1991; Ran, Z. et al. 2009. *Journal of Lightwave Technology*, 27(15), 3143–3149; Ran, Z. et al. 2011. *IEEE Sensors Journal*, 11(5), 1103–1106; Ran, Z. et al. 2011. In *21st International Conference on Optical Fiber Sensors (OFS-21)*, Ottawa, ON, Canada, Papers 7753-109, 7753-113, and 7753-114; Ran, Z. et al. 2011. *21st International Conference on Optical Fibre Sensors (OFS-21)* (pp. 775317–775317). International Society for Optics and Photonics.)

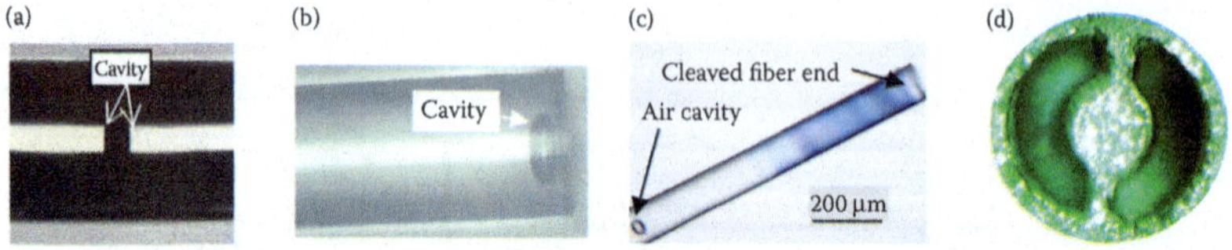

Figure 3.17 (a) Photograph of the self-enclosed FP cavity (strain sensor). (b) Photograph of the pressure sensor. (c) Photograph of the refractive and temperature sensor. (d) Top view of the accelerometer. (Rao, Y. J. and Ran, Z. L. 2011. *Laser Focus World*, 47(11), 71; Rao, Y. J. et al. 2007. *Optics Express*, 15(22), 14936–14941; Ran, Z. L. et al. 2008. *Optics Express*, 16(3), 2252–2263; Ran, Z. et al. 2013. *IEEE Sensors Journal*, 13(5), 1988–1991; Ran, Z. et al. 2009. *Journal of Lightwave Technology*, 27(15), 3143–3149; Ran, Z. et al. 2011. *IEEE Sensors Journal IEEE*, 11(5), 1103–1106; Ran, Z. et al. 2011. *21st International Conference on Optical Fiber Sensors (OFS-21)*, Ottawa, ON, Canada, Papers 7753-109, 7753-113, and 7753-114; Ran, Z. et al. 2011. *21st International Conference on Optical Fibre Sensors (OFS-21)* (pp. 775317–775317). International Society for Optics and Photonics.)

(see Figure 3.17d). This sensor is sensitive to the acceleration which direction is perpendicular to the fiber end face due to its double-clamped acceleration sensing structure. The sensor spectrum and response to acceleration are shown in Figure 3.21. Its sensitivity is ~1.8 rad/g.

3.2.3 Sapphire FFPI Sensor

Sapphire fiber (Al_2O_3) is a single-crystal material, whose melting point is 2040°C [10]. It is an ideal material to make a fiber-optic high-temperature sensor, but is difficult to be machined due to its extreme hardness. The sapphire fiber sensor structure, shown in Figure 3.22a, comprises a micro extrinsic air cavity and an intrinsic FP cavity. For comparison, a silica fiber-optic sensor with the same structure is also fabricated, as in Figure 3.22b. As the air cavity is not sensitive to temperature, the intrinsic FP cavity with a short section of fiber acts as a temperature-sensitive element. The sensor fabrication process consists of three basic steps: First, a series of 157-nm laser pulses are used to produce a circular hole on a sapphire fiber tip using a photo mask. Second, by splicing the sapphire fiber with a cleaved silica SMF end to enclose the hole, an air FP cavity was formed. Third, by cleaving the sapphire fiber near the air FP cavity using 157-nm laser, an FP temperature sensor is formed. The photo of the sapphire fiber sensor is shown in Figure 3.23.

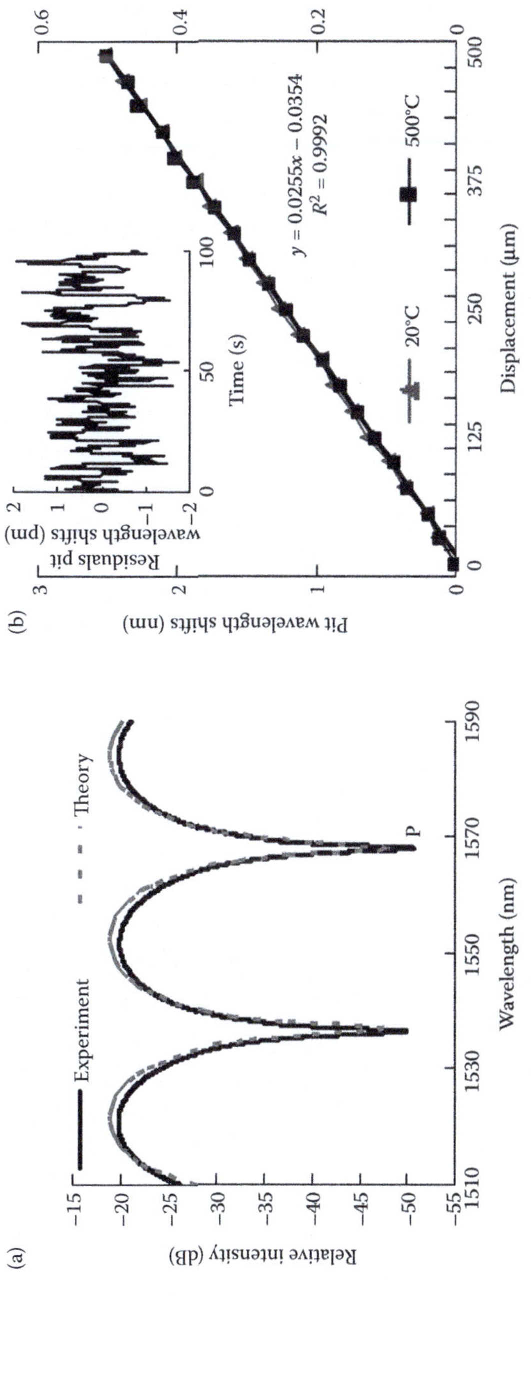

Figure 3.18 (a) Reflective spectrum of the etalon. (b) Static strain responses of the FPI. (From Ran, Z. L. et al. 2009. *Journal of Lightwave Technology*, 27(15), 3143–3149.)

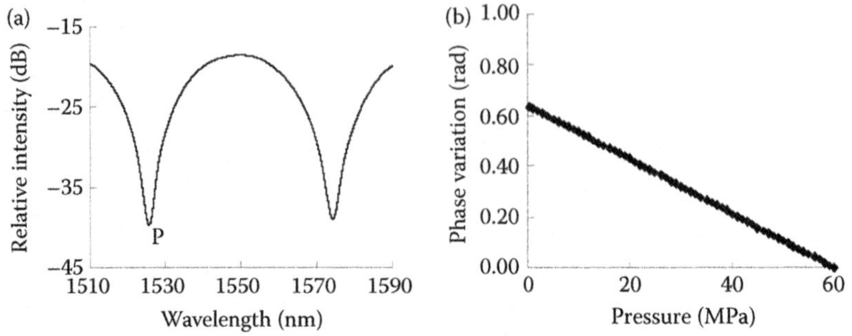

Figure 3.19 (a) Reflective spectrum of the pressure sensor. (b) Pressure responses of the sensor. (From Ran, Z. et al. 2011. *IEEE Sensors Journal*, 11(5), 1103–1106.)

The temperature characteristics of the sapphire fiber and silica fiber temperature sensors were both measured from 100°C to 1100°C. The spectra of the two sensors at different temperatures are shown in Figure 3.24a and b, respectively. It can be seen that the fringe contrast of the sapphire FP sensor is more stable than that of the silica FP sensor at high-temperature regions.

The temperature responses of the two types of sensors are shown in Figure 3.25. For the sapphire FP sensor, the pit wavelength shift at 1550 nm wavelength range is ~50.12 nm, corresponding to a sensitivity of ~50 pm/°C. It has good linearity of ~99.9%, while for the silica FP sensor, the wavelength shift is only ~8.86 nm, with a linearity of ~99.3%.

3.3 FIB Milling (Tapered Fibers)

FIB has been used in a number of fiber-based applications, including fabrication of long-period gratings, micromachining of fiber tips, and milling of side access holes in structured optical fibers. As a consequence of its applications in semiconductor technology, the FIB technique is mature and much more flexible than fs-laser processing, and dual-beam instruments combining a SEM and FIB are now commercially available. This method has the advantage of allowing micromachining with more well-defined milling profiles and a better surface finishing than what can be obtained with fs-laser technology with the conventional 800 nm wavelength, without the requirements of additional etching steps.

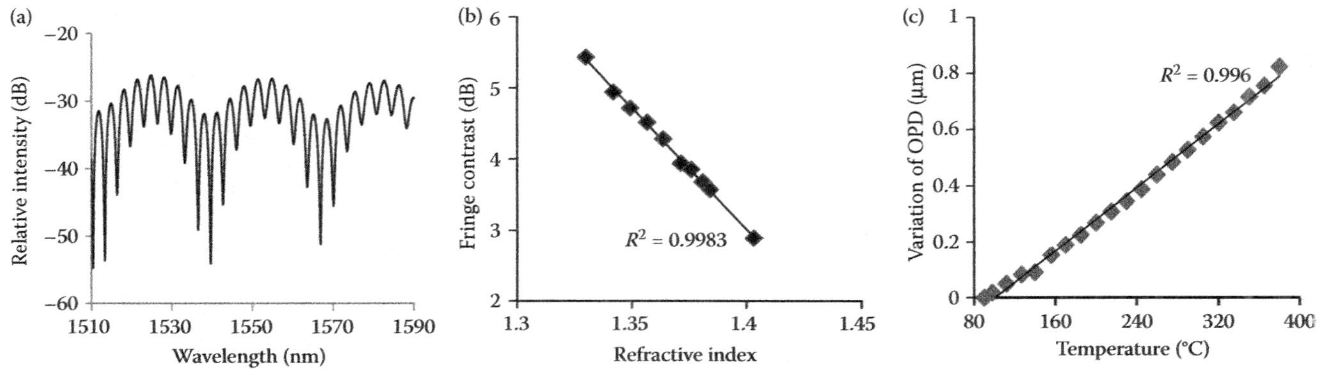

Figure 3.20 (a) Reflective spectrum of the temperature and refractive index sensor; (b) RI response; and (c) temperature response of the solid cavity (IFPI). (Ran, Z. L. et al. 2008. *Optics Express*, 16(3), 2252–2263; Ran, Z. et al. 2013. *IEEE Sensors Journal*, 13(5), 1988–1991.)

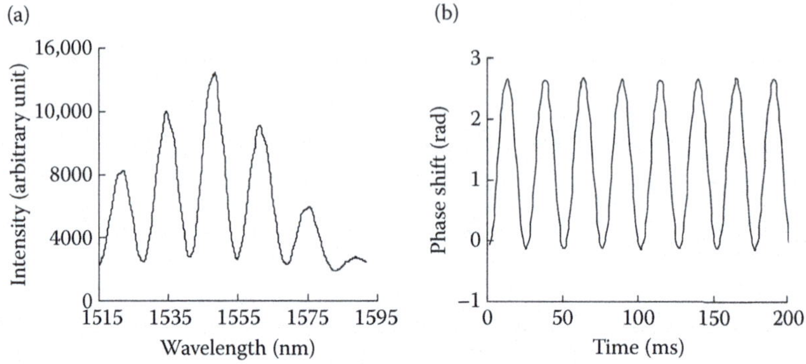

Figure 3.21 (a) Reflective spectrum of the FP sensor. (b) Measured phase shift of the sensor signal driven by a 40-Hz, 1.6-g peak-to-peak acceleration signal. (From Ran, Z. et al. 2011. *21st International Conference on Optical Fiber Sensors (OFS-21)*, Ottawa, ON, Canada, Papers 7753-109, 7753-113, and 7753-114.)

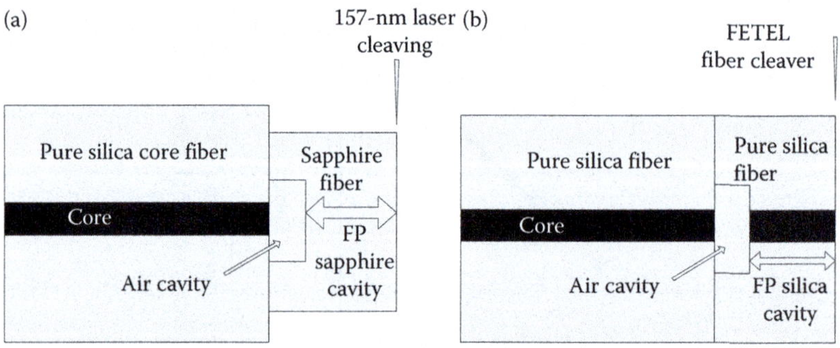

Figure 3.22 Schematic diagram of (a) sapphire fiber high-temperature sensor and (b) silica fiber sensor. (From Ran, Z. et al. 2011. *21st International Conference on Optical Fibre Sensors (OFS-21)* (pp. 775317–775317). International Society for Optics and Photonics.)

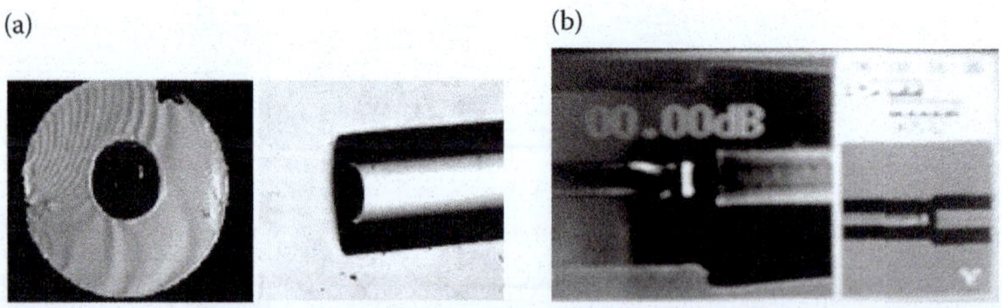

Figure 3.23 (a) Photo of micro-groove on fiber end. (b) Photo of the sapphire fiber temperature sensor in splicer. (From Ran, Z. et al. 2011. *21st International Conference on Optical Fibre Sensors (OFS-21)* (pp. 775317–775317). International Society for Optics and Photonics.)

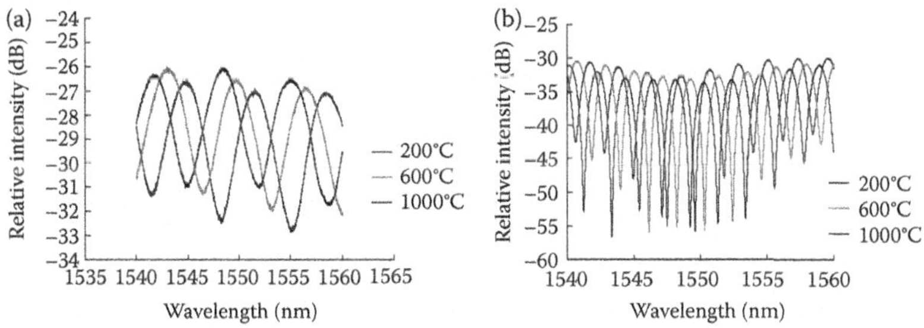

Figure 3.24 Reflected spectra of (a) sapphire FP sensor and (b) silica FP sensor. (From Ran, Z. et al. 2011. *21st International Conference on Optical Fibre Sensors (OFS-21)* (pp. 775317–775317). International Society for Optics and Photonics.)

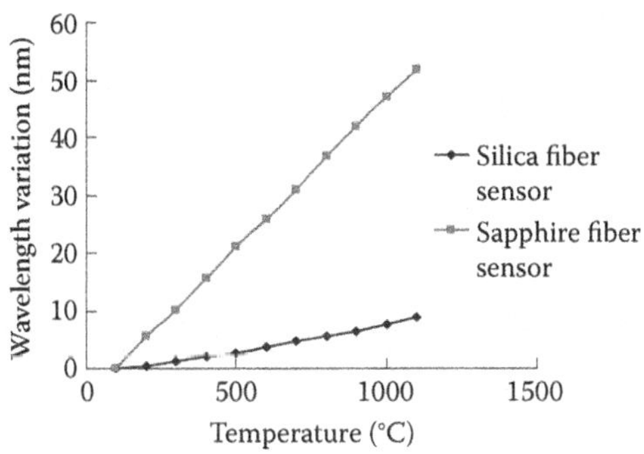

Figure 3.25 Temperature response of sapphire and silica FP sensors. (From Ran, Z. et al. 2011. *21st International Conference on Optical Fibre Sensors (OFS-21)* (pp. 775317–775317). International Society for Optics and Photonics.)

A Quanta 200 3D FIB with Ga^+ ion source is used to mill the fiber. The use of the FIB as a milling tool allows quick and direct patterning via a sputtering process. During sputtering, a portion of the ejected atoms or molecules are redeposited on the exposed region, making it difficult to control the amount of material that is removed. In order to avoid this redeposition process, the fiber is perpendicularly aligned with the ion beam and the milling process starts from the fiber side so that a pathway will be formed for the material ejection. Furthermore, a high incident angle (close to 90°) of the ion beam can speed up the milling rate significantly, since the FIB sputtering yield roughly increases with $1/\cos(\theta)$, where θ is the angle between the surface normal and the FIB direction [28].

During the ion bombardment process, there are excess charges cumulated on the surface of the material. These charges will cause random deflection of the incident ion beam and thus damage the milling profile. To avoid the charging effect, the fiber is fixed on the top of conductive carbon tape that is attached to the grounded sample holder. Moreover, a charge neutralization process with the aid of an electron beam is applied to counteract the charges. A conductive metal wire is also coiled as an anchor to prevent the fiber from vibrating. The micromachining is performed as close as possible to the anchored point.

To make a compact FP fiber sensor, a conventional SMF is used. The diameter of the fiber is tapered down to 32 μm to reduce the amount of surrounding silica that the FIB needs to penetrate before accessing the fiber core. This assists in reducing the milling time. To obtain a high milling rate, a high ion current of 20 nA is used at the beginning, which gives a relatively large beam spot size of 0.3–0.5 μm. As shown in Figure 3.26, the FIB milling starts from the side of the fiber to make an opening for the material ejection and reduce the material redeposition. After the speedy milling, a lower ion current (3 nA) is used to polish the two ends of the cavity with higher spatial resolution (down to 10 nm). A small additional tilt of the fiber (~3°) to the incident direction of the ion beam is applied during the polish milling, in order to get rid of the residual angle misalignment during milling on a tangent and obtain parallel end walls of the

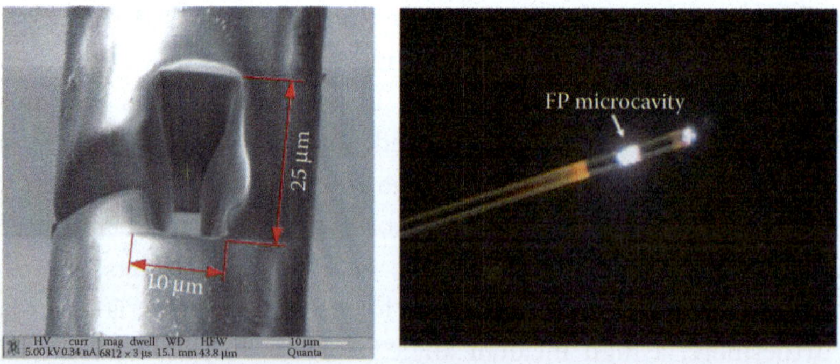

Figure 3.26 SEM image of the milled microcavity in the 32 μm in diameter fiber taper (left). Microscope image of the light guided in a ~500-μm-long fiber taper with the FP microcavity and the shattered fiber end (right). (From Yuan, W. et al. 2011. *Review of Scientific Instruments*, 82(7), 076103.)

microcavity. In this way, a 25-μm-long microcavity with a width of 10 μm is fabricated in the core of the fiber. The whole milling process takes around 20 min. The finely milled end facets of the microcavity can be seen in the enlarged SEM image in Figure 3.26. It is clearly seen that the inner sidewalls of the microcavity are not perfectly parallel, since 3~7° angle misalignment occurs due to the convergence of the FIB. This will not have a noticeable negative influence on the performance of the FP fiber sensor, since the surface quality of the end facets of the microcavity will have a much stronger influence on the interference pattern.

The microcavity is milled near the tip of the fiber taper and the taper end is shattered in order to reduce the Fresnel reflection, as shown in Figure 3.26. An experiment is carried out to test this microcavity as an FP fiber sensor. As shown in Figure 3.27, an FP interference pattern with an extinction ratio of about 15 dB and a pitch of about 37.4 nm is generated. Estimated with the wavelength of λ_{v1} and λ_{v2}, that is, $\lambda_{v1} = 1583.45$ nm and $\lambda_{v2} = 1546.05$ nm, and using the factory-defined refractive index value of the liquid at 25°C, that is, $n = 1.3$, a 25.274-μm FP cavity and a pitch of 37.28 nm are obtained from the relation $4\ \mathrm{nL} = m\lambda v$, where m is an odd integer, which is 83 for λ_{v1} and 85 for λ_{v2}. The whole spectrum shifts linearly toward shorter wavelength with increased temperature. An increased extinction ratio is also observed, which is due to the decrease in the refractive index of the liquid and the corresponding increase of the Fresnel reflection at the silica/liquid interface. By tracking λ_{v1} as shown in Figure 3.3, a total shift of 22.85 nm was introduced by a temperature increase of 38.8°C, which corresponds to an estimated decrease in the refractive index by 1.32×10^{-2}. This leads to a sensitivity of ~1731 nm/RIU. Taking into account the OSA resolution of 0.01 nm, the FP fiber sensor has a detection limit of ~5.78×10^{-6} RIU.

A fiber-integrated slot-type microresonator with strongly improved mirror reflectivity for precise refractive index sensing is shown in Figure 3.28. It is straightforward to implement and allows very simple and quick access to the actual sensing area. This resonator is formed by a vertical, micrometer-sized slot deeply milled into the core of a step-index fiber (SMF-28 type) using FIB. The resonator has its sides coated with a high-refractive index (HI) layer, which strongly

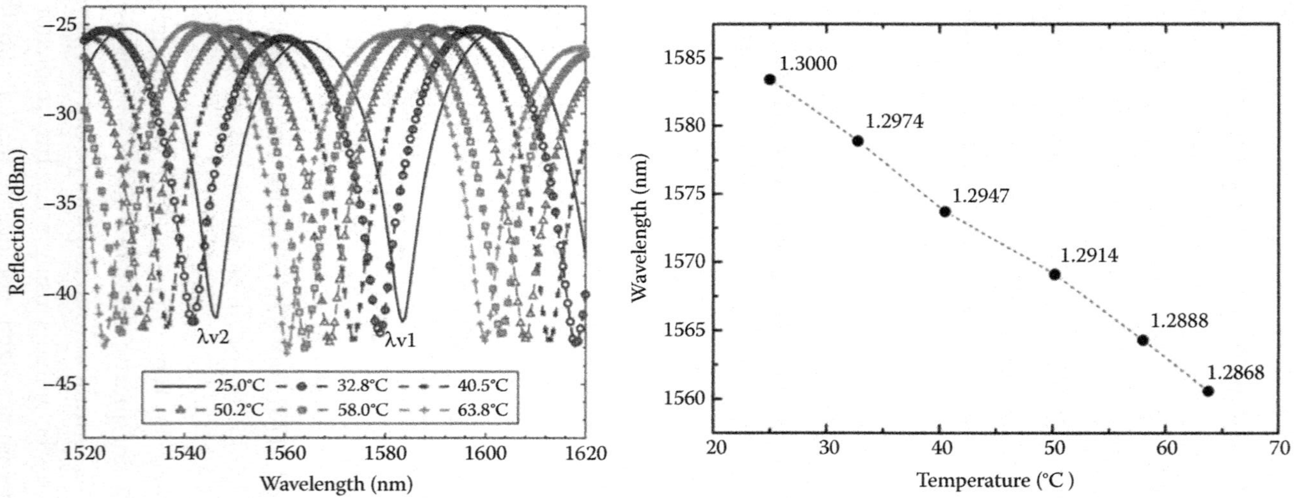

Figure 3.27 FP reflection spectrum for different temperatures of the refractive index liquid (left). Shift of λ_{vl} versus temperature; labeled numbers indicate the estimated refractive indices (right). (From Yuan, W. et al. 2011. *Review of Scientific Instruments*, 82(7), 076103.)

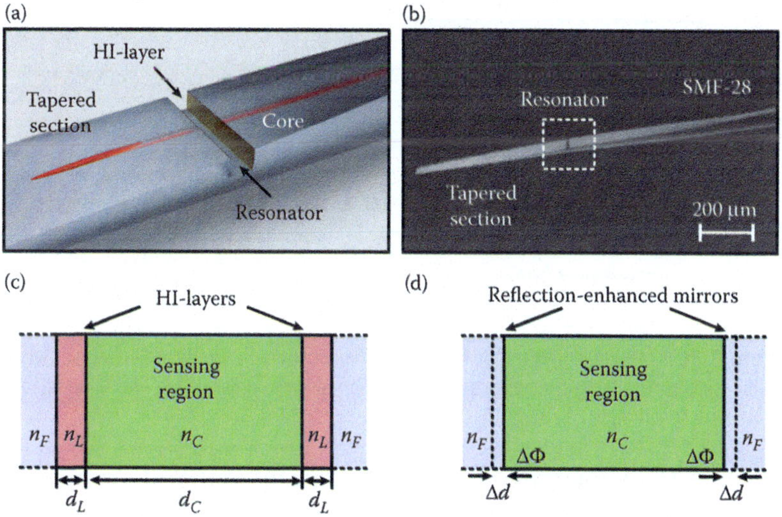

Figure 3.28 Reflectivity-enhanced in-fiber microresonator for precise refractive index sensing. (a) Schematic of the in-fiber microresonator with the deeply cut slot and the tapered section. (b) SEM of the fabricated fiber resonator (base fiber: SMF-28). (c) 1D transfer-matrix model of the resonator (green: slot representing the actual sensing area; pink: high-refractive index layers; light blue: silica). The parameters d_C and d_L indicate the extensions of cavity and layers; n_F, n_L, and n_C the refractive indices of fiber, layer, and slot. (d) Extended FP model with cavity extension Δd and additional phase shift $\Delta\phi$ (color code is identical to that in (c)). (From Wieduwilt, T. et al. 2014. *Optics Express*, 22(21), 25333–25346.)

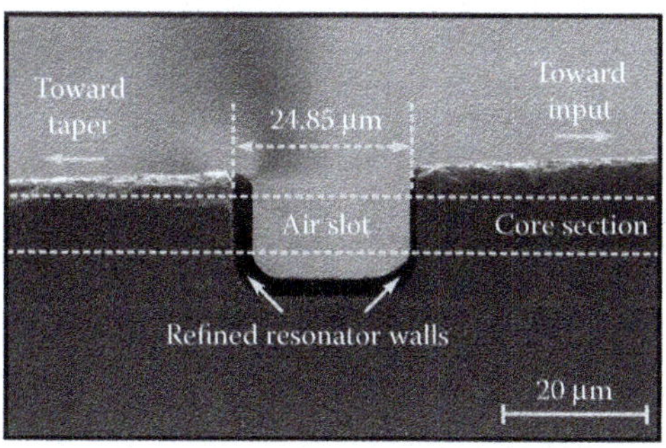

Figure 3.29 SEM of the focused-ion-beam milled and refined fiber-based microcavity (side view). The dashed yellow line represents the section of the guiding core. (From Wieduwilt, T. et al. 2014. *Optics Express*, 22(21), 25333–25346.)

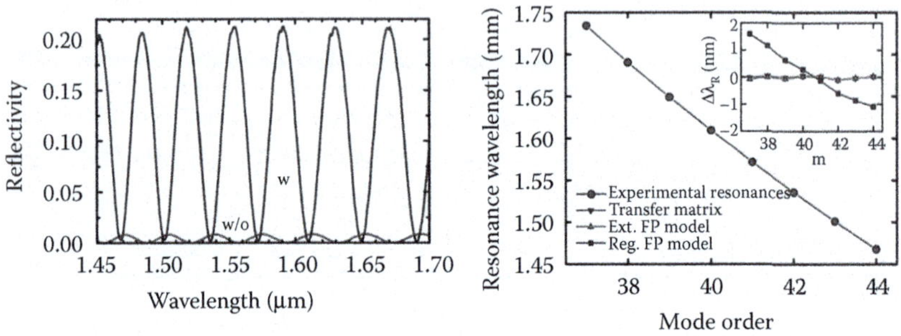

Figure 3.30 Measured spectral distribution of the normalized reflectivity (left). Distribution of the measurement reflection resonances as a function of mode order of the fiber-integrated FPR (right). (From Wieduwilt, T. et al. 2014. *Optics Express*, 22(21), 25333–25346.)

increases the amplitude of the reflected light and leads to a significantly enhanced fringe reflectivity difference ($\Delta RF = R_{max} - R_{min}$). The sensor allows measuring refractive indices within the range of silica glass, which is impossible using an uncoated resonator due to diminishing ΔRF.

The FP resonator was implemented by FIB milling (dual-beam system Lyra XMU [TESCAN]), a silica step-index fiber (SMF-28, Figure 3.29). Before milling, the fiber was polished at an angle of ~3° with respect to the fiber axis. The fiber was presputtered with platinum and electrically ground to avoid electrostatic charge accumulation on the fiber surface during the FIB processing. A two-step process was used: (i) high-current milling (10 nA) for cavity excavation and (ii) low-current processing (1 nA) to improve the parallelism and flatness of the cavity walls. The final cavity had a width and height of ~24.85 and ~18 μm, respectively.

The reflection spectra of the in-fiber FPR are measured, as shown in Figure 3.30. The spectral positions of the experimentally determined resonance wavelengths decrease almost purely linearly with increasing mode order (Figure 3.30), again resembling the typical behavior of an FPR.

3.4 Chemical Etching (Doped Fibers)

3.4.1 Strain Sensor Fabricated by Chemical Etching

Chemical etching micromachining technique could be used to fabricate FFP sensor in a special-doped fiber, which is based on the much

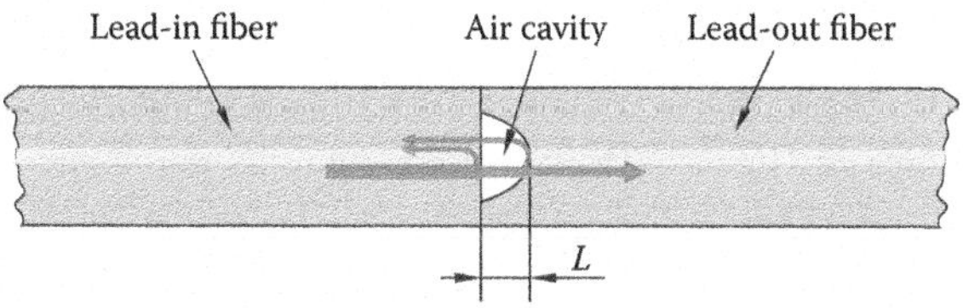

Figure 3.31 In-line optical fiber strain sensor. (From Cibula, E. and Donlagic, D. 2007. *Optics Express*, 15(14), 8719–8730.)

higher etching rate of doped silica when compared to pure silica. Such an FP sensor is generally used as a strain sensor. The sensor design is shown in Figure 3.31. A short air cavity between the lead-in and lead-out SMF acts as an FPI, where light waves, reflected from both cavity surfaces, interfere.

The fabrication of a short-cavity strain sensor includes three steps: concave cavity formation at the tip of the lead-out fiber, splicing the fibers to perform an in-line sensor, and, finally, tuning of the sensor's length to achieve the desired operating point.

First the short segment of the standard 62.5/125 gradient index MMF is attached to a standard SMF by splicing and cleaving at the desired distance from the splice, in our case 25 μm, as shown in Figure 3.32a. The appropriate tools were used to achieve repeatable cleaves with 1 μm resolution [30]. This initial 25 μm distance is shorter than the target cavity length, as the fusion splicing process

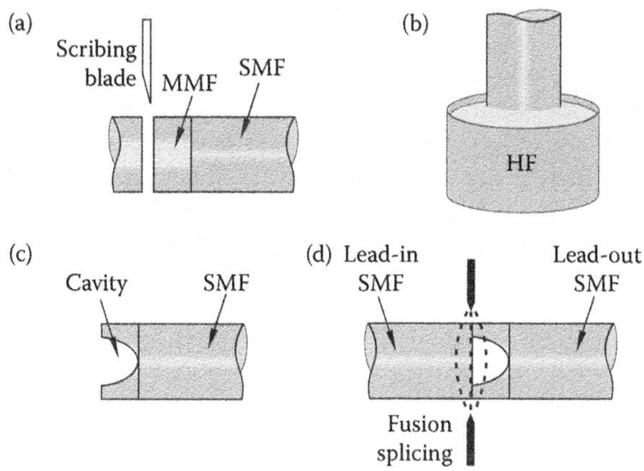

Figure 3.32 Strain sensor fabrication procedure. (From Cibula, E. and Đonlagić, D. 2005. *Applied Optics*, 44(14), 2736–2744.)

causes the elongation of the cavity. Next, the prepared fiber tip is immersed in 40% HF acid at room temperature (Figure 3.32b), where the core of the MMF is etched away in the form of a concave cavity, as illustrated in Figure 3.32c. To assure high reflectivity of the structure, it is essential to terminate the etching process at the moment the acid solution reaches the MMF–SMF boundary at the center of the fiber. They adopted the etching procedure described in Reference 31 where the reflectivity is continuously measured during etching using an appropriate optical system. The fiber is removed from the acid at the moment when the reflectivity reaches maximum value. In the following step, the prepared fiber is spliced to another flat-cleaved SMF, thus forming the sensor, as shown in Figure 3.32d.

Splicing is performed using a standard fusion splicer (Ericsson FSU 925 PM-A). While the first splice between the MMF and the SMF is achieved using standard parameters for MMF–MMF splicing, the second splice required careful adjustment of fusion parameters to ensure high reflectivity and high interference contrast simultaneously with high tensile strength. An enlarged photograph of the produced cavity is shown in Figure 3.33. The reflected spectrum of the tuned sensor is shown in Figure 3.33.

3.4.2 Pressure Sensor Fabricated by Chemical Etching

An all-glass miniature (Ø125 μm) fiber-optic pressure sensor design is appropriate for high-volume manufacturing. The fabrication process is based on the chemical etching of specially designed silica

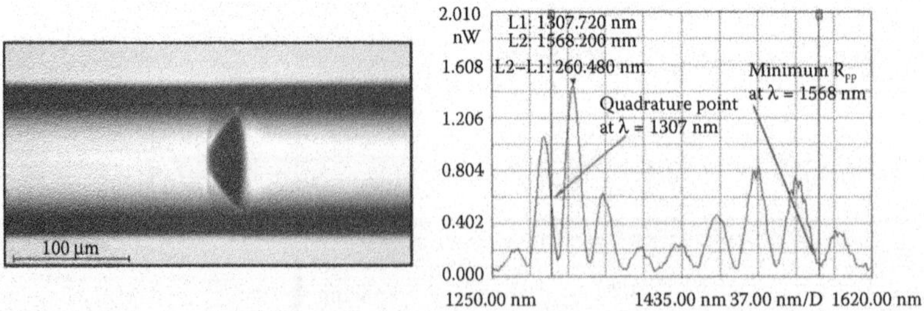

Figure 3.33 Enlarged photograph of the strain sensor (left). Reflected spectrum of tuned sensor (right). (From Cibula, E. and Đonlagić, D. 2005. *Applied Optics*, 44(14), 2736–2744.)

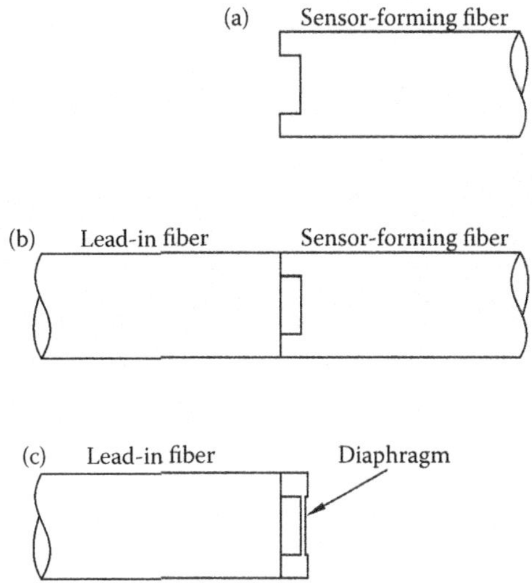

Figure 3.34 Pressure sensor simplified fabrication process is performed over only three steps: (a) formation of the cavity, (b) fusion splicing with the lead-in fiber, and (c) creation of the diaphragm. (From Cibula, E. et al. 2009. *Optics Express*, 17(7), 5098–5106.)

optical fiber and involves a low number of critical production operations. The presented sensor design can be used with either single-mode or multimode lead-in fiber and is compatible with various types of available signal processing techniques. The practical assembly of the sensor is achieved over only three steps, which are summarized in Figure 3.34.

In the first step, a cavity is micromachined at the tip of a cleaved optical fiber with well-defined depth and diameter. The cavity is created by the selective chemical etching of a custom-designed, sensor-forming optical fiber. This sensor-forming fiber has a germanium-doped step-index core with a diameter of 90 μm and a refractive index difference of about 0.7% (corresponding to 7.42% mol of GeO_2). When this fiber is exposed to 40% HF at 25°C, the doped region etches at a rate of about 2.8 μm/min, while the pure silica cladding etches at a rate of about 1 μm/min. The higher etching rate of the doped core region results in a cavity formation at the tip of the fiber, as shown in Figure 3.34a. The cavity depth is simply controlled by etching time. Depending on the desired cavity depth, the sensor-forming fiber is drawn to a larger initial diameter in such a way as to achieve a standard 125-μm fiber outer diameter after completion of the chemical etching phase. For example, when a target cavity depth

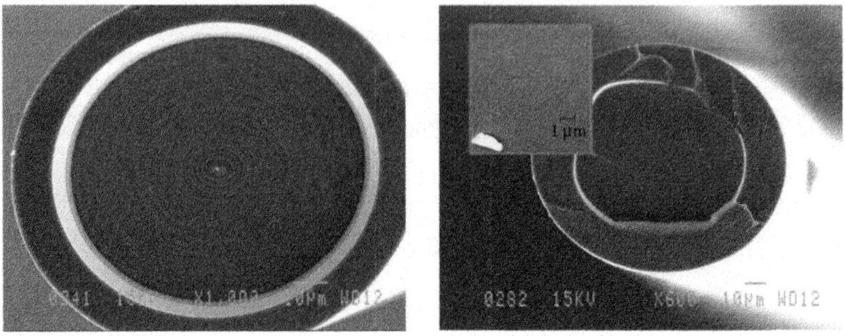

Figure 3.35 SEM photos of the etched cavity: (a) before fire-polishing and (b) after fire-polishing. (From Cibula, E. et al. 2009. *Optics Express*, 17(7), 5098–5106.)

of 12 μm was selected, the starting (initial) sensor-fiber diameter corresponded to 138 μm and the etching took 6.5 min in 40% HF at 25°C. Various cavity depths were produced by using different etching times and different initial outer diameters of the sensor-forming fiber.

In the second step, the etched sensor-forming fiber is fusion-spliced to the lead-in fiber, as shown in Figure 3.34b. Any desired type of lead-in fiber can be used depending on the signal processing technique. Fusion splicing is also used to fire-polish the inner surface of the cavity that becomes rough after chemical etching due to ring MCVD layers. The cavity end-surface qualities before and after fire-polishing based on optimized fusion splicing procedure are shown in Figure 3.35.

In the third fabrication step, the sensor-forming fiber is cleaved near the fusion splice with the help of visual inspection under a microscope [10] and is polished to form a pressure-sensitive diaphragm (Figure 3.34c).

Figure 3.36 shows a picture of the produced sensor with a typical 550 nm/bar sensitivity, obtained using a SEM. The response characteristic at such high pressures becomes highly nonlinear, as shown in Figure 3.36.

3.4.3 Vibration Sensor Fabricated by Combination of Chemical Etching and FIB

FIB technology is combined with chemical etching of specifically designed fibers to create FPIs. Hydrofluoric acid is used to

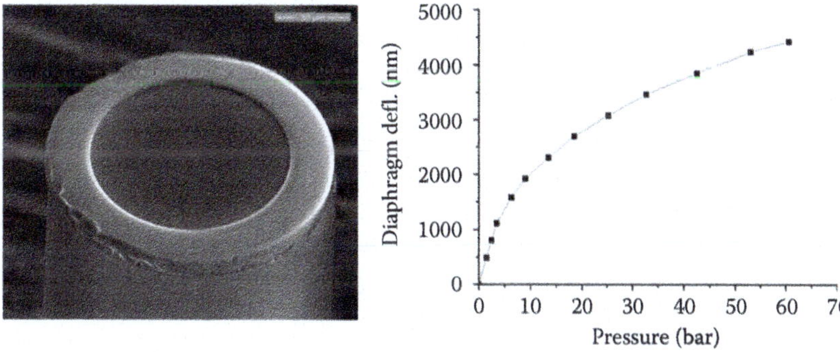

Figure 3.36 SEM photo of a typical sensor with a sensitivity of ~550 nm/bar (×2000, left). Deflection of sensor diaphragm versus pressure at high pressures showing nonlinear behavior (sensor sensitivity 550 nm/bar, right). (From Cibula, E. et al. 2009. *Optics Express*, 17(7), 5098–5106.)

etch special fibers and create microwires with diameters of 15 μm. These microwires are then milled with an FIB to create two different structures: an indented FP structure and a cantilever FP structure that are characterized in terms of temperature [33].

The fabrication of the FP structures can be divided into two steps: the fabrication of the microwire by chemical etching micromachining, and the milling of a gap in the microwire with FIB technology. This two-step process allows for the much faster fabrication of microstructures than solely using FIB on standard fiber. Accessing the light guiding region with FIB would take too long on a standard fiber, and the structures would be very poorly defined, due to the high aspect ratio necessary.

This micromachining technique is based on the much higher etching rate of phosphorus pentoxide-doped silica when compared to pure silica. This way, structure-forming fibers (SFFs) can be engineered with pure silica regions and P_2O_5-doped regions so that, after etching, only the pure silica regions remain, leaving just the desired microstructure [34,35]. This technique is used to create microwires, which are then further post-processed using FIB technology. After splicing SFF to SMF, the SFF is cleaved to the desired length (see Figure 3.37a–f). To prevent etching from the top of the fiber, an additional short section of a coreless all-silica multimode fiber (cMMF) is spliced to the top of the SFF. The SFF was cleaved using an ultrasonic YORK FK 11 cleaver set at a tensile strength of 2 N. The splicing was performed

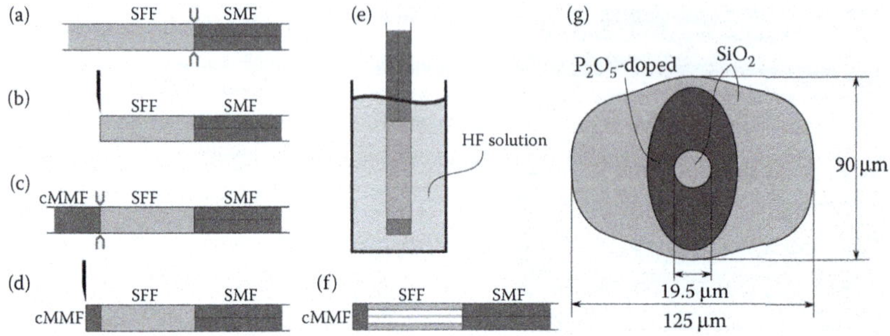

Figure 3.37 Microwire fabrication process: (a) SMF–SFF fusion splicing; (b) cleaving to desired length; (c) SFF–cMMF fusion splicing; (d) cMMF cleaving (30–40 μm); (e) etching; (f) final structure; (g) SEM micrograph of etched microwire. (From André, R. M. et al. 2014. *Optics Express*, 22(11), 13102–13108.)

by a filament fusion splicer (Vytran FFS 2000) that led to splices with losses below 0.2 dB [35].

The whole structure is then placed inside a HF solution with 40% concentration. Initially, only pure silica is in contact with the solution and, consequently, the whole structure is etched uniformly, but when the outer silica shell is etched away and the acid comes into contact with the doped region, preferential etching of the P_2O_5-doped silica occurs. The P_2O_5 concentration of the SFF is about 8.5 mol%, which means that the etching rate of the P_2O_5-doped region is about 30 times higher than the etching rate of pure silica. Etching in 40% HF at room temperature (~25°C) with no stirring leads to etching rates of 1 μm/min for pure silica and 31 μm/min for the P_2O_5-doped region. The process was concluded by rinsing the structures in distilled water. The total etching times depend on the desired microwire diameter and the external temperature and can range from 15 to 20 min. The structure that remains after chemical etching consists of a microwire with a diameter of 15 μm, aligned with the single-mode lead-in fiber core and two side support beams that, due to the complete misalignment with the SMF core, do not guide light (see Figure 3.37f). These side support beams give the microwire protection and help the whole structure retain its form. Even though the microwire, when in air, supports several modes after being etched, practically only one mode is launched by the SMF in the current configuration. This required special care in structure design as described in detail in Reference 35. The guiding losses for the microwires are below 0.4 dB for diameters of 15 μm.

After the microwire is created, an FIB is used to mill the microwire and create two different FP structures. Before FIB milling, the microwires were sputter-coated with a thin tantalum film (ca. 50 nm). This is necessary to avoid charging during electron beam and ion beam operation of the fiber, as silica is nonconductive. The charging will affect the milling because it will cause the ion beam to drift from its intended spot position, effectively reducing the resolution and quality of the milled structures. In the milling of these structures, an ion current of approximately 1 nA is used for a primary coarse milling of the cavities. After this, a polishing is performed using a much smaller current of 100–300 pA. The currents were adjusted so that the primary milling times did not exceed 1 h and the secondary polishing times did not exceed 20 min to avoid charging and consequent drift effects. The surface quality is rough after the primary milling due to the high current employed and also due to the redeposition of some of the milled material but, after the polishing, the surface roughness greatly decreases. Flat, parallel walls can be obtained because the aspect ratio of the milled structure is not high [36]. The first structure milled consists of an indentation in the microwire (see Figure 3.38a). The reflections at both silica-to-air interfaces (signaled in Figure 3.38a) result in a low-finesse FP cavity. The cavity has a length of approximately 167 μm.

The second structure is similar save that a whole section of the microwire is removed instead of just a half-cylinder section (see Figure 3.38b). This results in a completely cleaved microwire that is suspended from the fiber top side. The microwire stays in place due to the side support beams that still remain after the milling process. This structure also behaves as an FP cavity, being that the reflecting interfaces are the fiber top and the silica-to-air interface at the air gap signaled in Figure 3.38b. In this case, the cavity has a much greater length of approx. 1025 μm.

In the cantilever FP structure, the microwire is solely suspended by one of its ends as opposed to the indented FP structure where both ends of the microwire are fixed. This suspended microwire has freedom to move relative to the bulk input fiber, allowing for aligned and misaligned positions. Using this property, it is possible to apply this structure as a vibration sensor. The structure was attached to an

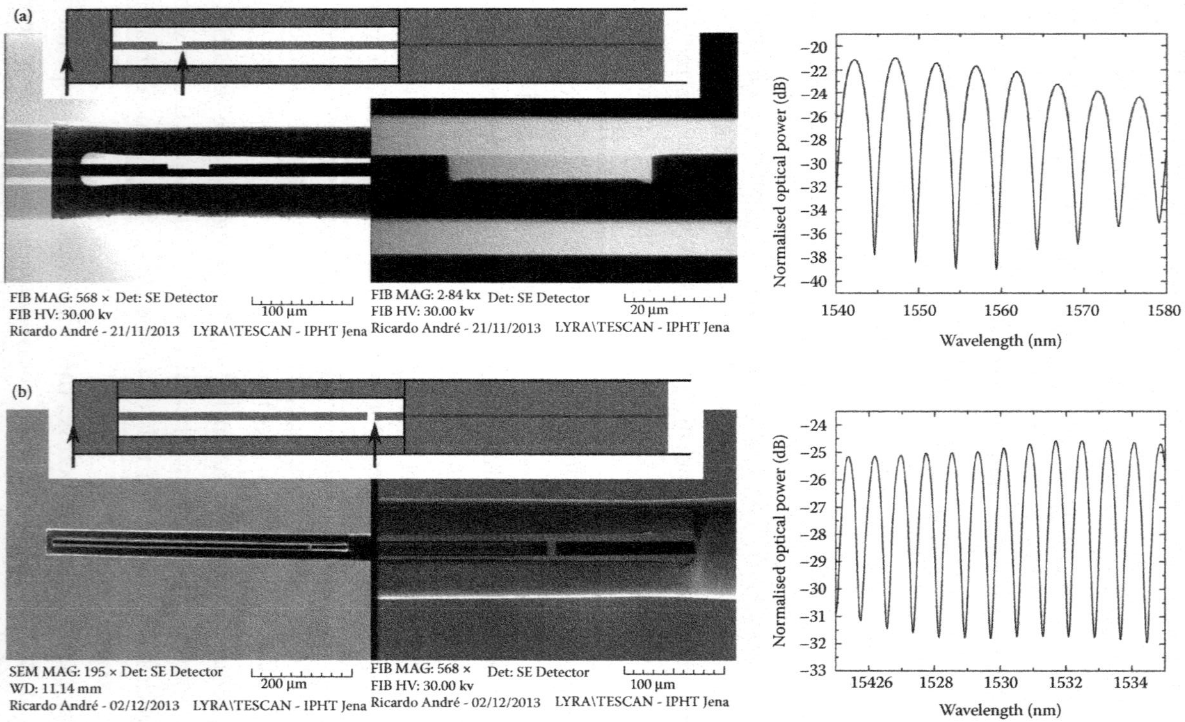

Figure 3.38 (a) Indented FP cavity SEM micrographs and related optical reflection spectrum. (b) Cantilever FP structure micrograph and related optical reflection spectrum. (From André, R. M. et al. 2014. *Optics Express*, 22(11), 13102–13108.)

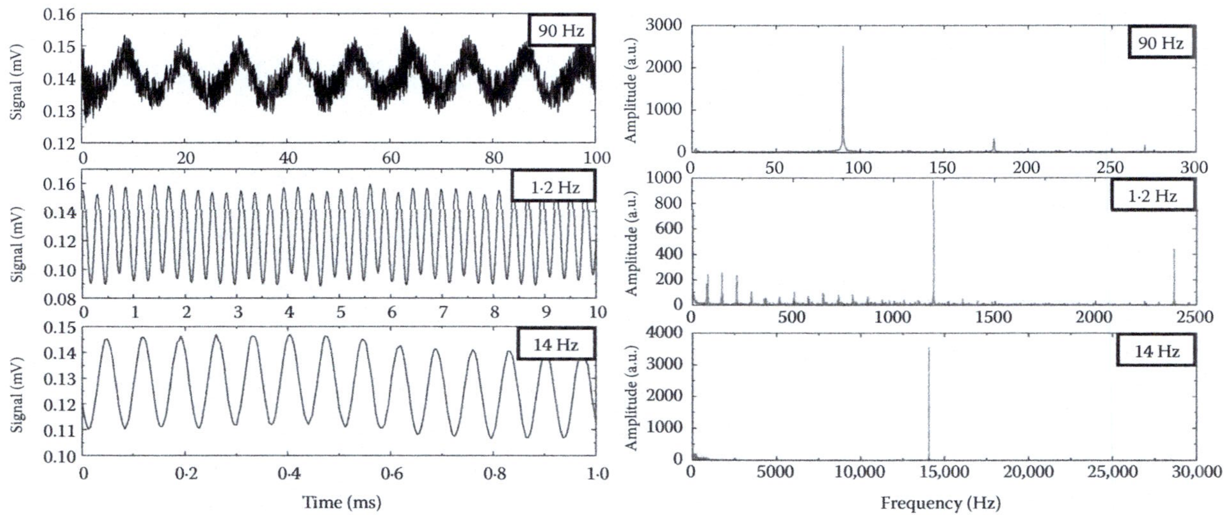

Figure 3.39 FP cantilever structure: time responses (left) and related fast Fourier transforms (right) when an external frequency is applied. (From André, R. M. et al. 2014. *Optics Express*, 22(11), 13102–13108.)

acoustic vibrating system that produced a vibration frequency in the range from 1 Hz to 40 kHz, as shown in Figure 3.39.

3.5 Summary

Different fabrication methods were compared as follows. They could be used to realize all kinds of micro FP sensors with their particular advantages and disadvantages.

METHODS	FABRICATION	SPEED	MASS PRODUCTION ABILITY	SENSOR TYPE	COST
Fs laser	Sweep ablation	Low	Low	Strain, RI, pressure	High
UV laser	Sweep or mask ablation	High	Comparable	Strain, RI, pressure, accelerator	High
FIB	Sweep milling	Low	Low	RI, vibration	High
Chemical etching	Batch etching	High	Good	Strain, pressure, vibration	Low

References

1. Rao, Y. J. 2006. Recent progress in fiber-optic extrinsic Fabry–Perot interferometric sensors. *Optical Fiber Technology*, 12(3), 227–237.
2. Petuchowski, S. J., Giallorenzi, T. G., and Sheem, S. K. 1981. A sensitive fiber-optic Fabry–Perot interferometer. *IEEE Journal of Quantum Electronics*, 17, 2168–2170.
3. Lee, C. E. and Taylor, H. F. 1988. Interferometric optical fibre sensors using internal mirrors. *Electronics Letters*, 24(4), 193–194.
4. Wan, X. and Taylor, H. F. 2002. Intrinsic fiber Fabry–Perot temperature sensor with fiber Bragg grating mirrors. *Optics Letters*, 27(16), 1388–1390.
5. Kersey, A. D., Jackson, D. A., and Corke, M. 1983. A simple fibre Fabry–Perot sensor. *Optics Communications*, 45(2), 71–74.
6. Jiang, L. and Tsai, H. L. 2006. Plasma modeling for ultrashort pulse laser ablation of dielectrics. *Journal of Applied Physics*, 100(2), 023116.
7. Davis, K. M., Miura, K., Sugimoto, N., and Hirao, K. 1996. Writing waveguides in glass with a femtosecond laser. *Optics Letters*, 21(21), 1729–1731.
8. Szameit, A., Bloemer, D., Burghoff, J., Pertsch, T., Nolte, S., and Tünnermann, A. 2006. Hexagonal waveguide arrays written with fs-laser pulses. *Applied Physics B*, 82(4), 507–512.
9. Sun, H., He, F., Zhou, Z., Cheng, Y., Xu, Z., Sugioka, K., and Midorikawa, K. 2007. Fabrication of microfluidic optical waveguides on glass chips with femtosecond laser pulses. *Optics Letters*, 32(11), 1536–1538.

10. Glezer, E. N., Milosavljevic, M., Huang, L., Finlay, R. J., Her, T. H., Callan, J. P., and Mazur, E. 1996. Three-dimensional optical storage inside transparent materials. *Optics Letters*, 21(24), 2023–2025.
11. Rao, Y. J., Deng, M., Duan, D. W., Yang, X. C., Zhu, T., and Cheng, G. H. 2007. Micro Fabry–Perot interferometers in silica fibers machined by femtosecond laser. *Optics Express*, 15(21), 14123–14128.
12. Kondo, Y., Nouchi, K., Mitsuyu, T., Watanabe, M., Kazansky, P. G., and Hirao, K. 1999. Fabrication of long-period fiber gratings by focused irradiation of infrared femtosecond laser pulses. *Optics Letters*, 24(10), 646–648.
13. Wei, T., Han, Y., Tsai, H. L., and Xiao, H. 2008. Miniaturized fiber inline Fabry–Perot interferometer fabricated with a femtosecond laser. *Optics Letters*, 33(6), 536–538.
14. Wei, T., Han, Y., Li, Y., Tsai, H. L., and Xiao, H. 2008. Temperature-insensitive miniaturized fiber inline Fabry–Perot interferometer for highly sensitive refractive index measurement. *Optics Express*, 16(8), 5764–5769.
15. Liao, C. R., Hu, T. Y., and Wang, D. N. 2012. Optical fiber Fabry–Perot interferometer cavity fabricated by femtosecond laser micromachining and fusion splicing for refractive index sensing. *Optics Express*, 20(20), 22813–22818.
16. Zhang, Y., Yuan, L., Lan, X., Kaur, A., Huang, J., and Xiao, H. 2013. High-temperature fiber-optic Fabry–Perot interferometric pressure sensor fabricated by femtosecond laser. *Optics Letters*, 38(22), 4609–4612.
17. Huang, J., Lan, X., Luo, M., and Xiao, H. 2014. Spatially continuous distributed fiber optic sensing using optical carrier based microwave interferometry. *Optics Express*, 22(15), 18757–18769.
18. Wang, W., Pang, F., Chen, N., Zhang, X., Lan, L., Ding, D., and Wang, T. 2010, December. Fiber optic intrinsic Fabry–Perot temperature sensor fabricated by femtosecond lasers. In *Asia Communications and Photonics Conference and Exhibition* (pp. 79900P–79900P), Shanghai, China. International Society for Optics and Photonics.
19. Ran, Z. L., Rao, Y. J., Deng, H. Y., and Liao, X. 2007. Miniature in-line photonic crystal fiber etalon fabricated by 157 nm laser micromachining. *Optics Letters*, 32(21), 3071–3073.
20. Rao, Y. J. and Ran, Z. L. 2011. Fibers for sensing-fiber-optic Fabry–Perot sensors take the strain under high temperature. *Laser Focus World*, 47(11), 71.
21. Rao, Y. J., Ran, Z. L., Liao, X., and Deng, H. Y. 2007. Hybrid LPFG/MEFPI sensor for simultaneous measurement of high-temperature and strain. *Optics Express*, 15(22), 14936–14941.
22. Ran, Z. L., Rao, Y. J., Liu, W. J., Liao, X., and Chiang, K. S. 2008. Laser-micromachined Fabry–Perot optical fiber tip sensor for high-resolution temperature-independent measurement of refractive index. *Optics Express*, 16(3), 2252–2263.

23. Ran, Z., Li, C., Zuo, H., and Chen, Y. 2013. Laser-machined cascaded micro cavities for simultaneous measurement of dual parameters under high temperature. *IEEE Sensors Journal*, 13(5), 1988–1991.
24. Ran, Z. L., Rao, Y. J., Liao, X., and Deng, H. Y. 2009. Self-enclosed all-fiber in-line etalon strain sensor micromachined by 157-nm laser pulses. *Journal of Lightwave Technology*, 27(15), 3143–3149.
25. Ran, Z., Liu, Z., Rao, Y., Xu, F., Sun, D., Yu, X., Xu, B., and Zhang, J. 2011. Miniature fiber-optic tip high pressure sensors micromachined by 157 nm laser. *IEEE Sensors Journal*, 11(5), 1103–1106.
26. Ran, Z., Lu, E., Rao, Y., Ni, M., Peng, F., and Zeng, D. 2011. Fiber optic Fabry–Perot interferometer tip accelerometer fabricated by laser-micromachining. In *21st International Conference on Optical Fiber Sensors (OFS-21)*, Ottawa, ON, Canada, Papers 7753–114.
27. Ran, Z., Lu, E., Rao, Y., Peng, F., and Liu, Z. 2011. 1100°C fiber-optic high-temperature Fabry–Perot sensors fabricated by laser-micromachining. In *21st International Conference on Optical Fibre Sensors* (OFS-21) (pp. 775317–775317), Ottawa, ON, Canada. International Society for Optics and Photonics.
28. Yuan, W., Wang, F., Savenko, A., Petersen, D. H., and Bang, O. 2011. Note: Optical fiber milled by focused ion beam and its application for Fabry–Pérot refractive index sensor. *Review of Scientific Instruments*, 82(7), 076103.
29. Wieduwilt, T., Dellith, J., Talkenberg, F., Bartelt, H., and Schmidt, M. A. 2014. Reflectivity enhanced refractive index sensor based on a fiber-integrated Fabry–Perot microresonator. *Optics Express*, 22(21), 25333–25346.
30. Cibula, E. and Donlagic, D. 2007. In-line short cavity Fabry–Perot strain sensor for quasi distributed measurement utilizing standard OTDR. *Optics Express*, 15(14), 8719–8730.
31. Cibula, E. and Đonlagić, D. 2005. Miniature fiber-optic pressure sensor with a polymer diaphragm. *Applied Optics*, 44(14), 2736–2744.
32. Cibula, E. Pevec, S., Lenardic, B., Pinet, E., and Donlagic, D. 2009. Miniature all-glass robust pressure sensor. *Optics Express*, 17(7), 5098–5106.
33. André, R. M., Pevec, S., Becker, M., Dellith, J., Rothhardt, M., Marques, M. B., Donlagic, D., Bartelt, H., and Frazão, O. 2014. Focused ion beam post-processing of optical fiber Fabry–Perot cavities for sensing applications. *Optics Express*, 22(11), 13102–13108.
34. Pevec, S. and Donlagic, D. 2011. All-fiber, long-active-length Fabry–Perot strain sensor. *Optics Express*, 19(16), 15641–15651.
35. Pevec, S. and Donlagic, D. 2012. Miniature micro-wire based optical fiber-field access device. *Optics Express*, 20(25), 27874–27887.
36. Heo, J. S., Lee, J. J., and Lim, J. O. 2003. A micro total reflective extrinsic Fabry–Perot interferometric fiber optic pressure sensor for medical application. *International Journal of Modern Physics B*, 17(08n09), 1199–1204.

4

Physical and Biochemical Sensors Based on FFPIs

Fiber-optic sensors have been widely used in a number of fields, ranging from distributed sensing of strain at kilometer scale and biomolecule detection at micrometer or even nanometer scale. For FFP sensors, they have mainly been used with miniature size for sensing of both physical and biochemical parameters.

4.1 Physical Sensors

Physical parameters, including temperature, strain, displacement, force, pressure, acoustic or ultrasonic waves, and electric and magnetic fields, were detected for smart structure and infrastructure health monitoring, oil/gas and electric power industry, etc.

4.1.1 Temperature Sensors

Temperature detection is required in a lot of industrial applications, in order to save energy, or to monitor the production process during which temperature may influence much. Conventional temperature detection methods include the thermocouple, heat resistance, and infrared radiation detection. They may be influenced by ambient electromagnetic interference or may not be available for applications in harsh environments. Fiber-optic temperature sensors can withstand temperatures up to 1000°C and have the advantages of immune to electromagnetic interference, and environmental vibrations.

Fiber-optic FPIs were first embedded in composite materials for temperature sensing, demonstrated by H. Taylor and coworkers [1,2]. The measurement range was 20–200°C when the FFPI was embedded in a graphite–epoxy composite material. They also demonstrated that the temperature sensitivity can be enhanced by 2.9 times when the FFPI was embedded into stainless-steel tubes.

One of the most important applications of FFPI sensors is for high-temperature measurement, thanks to the inherent high melting point of silica. FFPI sensors based on the common Ge-doped optical fibers can be used under a temperature of about 600°C [3]. By using special optical fibers such as PCFs or sapphire fibers, the measurement range can be extended up to 1200°C or even above 1500°C [4]. There were numerous papers related to fiber-optic high-temperature sensing, and another bunch of papers for strain or pressure sensing under high-temperature environment.

Figure 4.1 shows the high-temperature sensing performance of an FFPI sensor based on fusion splicing a section of solid-core PCF to SMF. A temperature range of –20°C to 1200°C was measured with good linearity and good repeatability [5].

The most recent report on FFPI temperature sensors described a sourceless high-temperature sensor [6], in which the thermal radiation, with a broad band, of the measurement environment was employed as light source for the detection of FP interference and no external light source was required. This is a smart idea that significantly simplifies the experimental setup and also reduces the cost. By employing the interference from a sapphire wafer, the fringe contrast of the reflective spectrum was good and the temperature can be measured up to 1593°C, as shown in Figure 4.2.

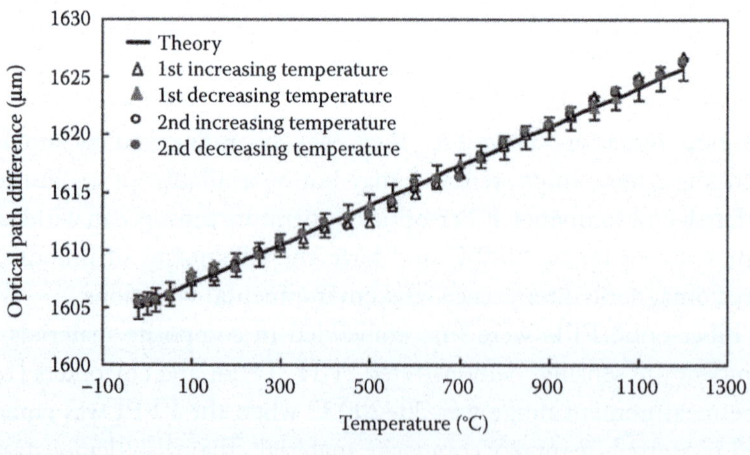

Figure 4.1 High-temperature sensing performance of PCF-FFPI.

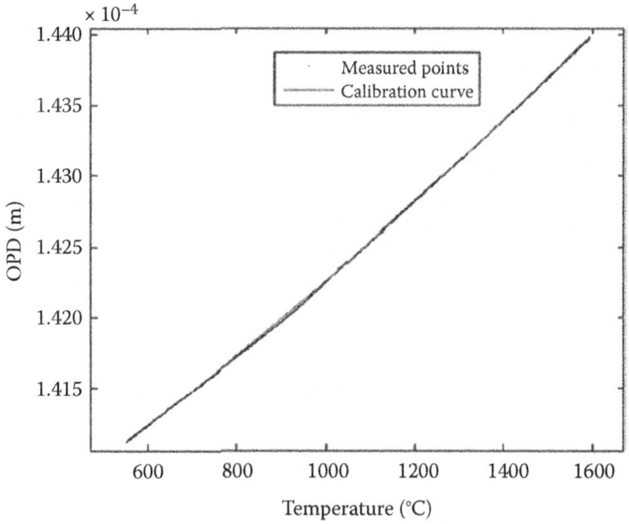

Figure 4.2 Sourceless high-temperature FFPI sensor based on sapphire wafer.

As described in Chapter 2, FFPI can be formed with different kinds of optical fibers or fabricated by various micromachining technologies. Theoretically speaking, all of these FFPIs can be used for temperature sensing, as long as there is no cross talk between temperature and other parameters like strain. For example, Lee et al. [7] performed high-temperature sensing by using an FFPI based on dielectric films coated on the end face of optical fiber layer-by-layer. The coating was annealed under 800°C and was still smooth. Temperature sensing between 250°C and 750°C was demonstrated by wavelength shift detection.

Li et al. [8] developed an ultrahigh-sensitive FFPI temperature sensor based on a graphene diaphragm. The graphene film was grown by chemical vapor deposition (CVD) method and transferred onto a ferrule into which the fiber was inserted. The fiber end and the graphene film formed an FFPI, and the wavelength shift of the interference fringes was measured. A high sensitivity of 1.87 nm/°C was achieved, about two orders of magnitude higher than that of the conventional intrinsic FFPIs. The FFPI can be used for high-temperature sensing of up to 1000°C. There were also FFPIs based on other materials like silicon, which has the advantage of temperature sensing at high frequency of up to 2 kHz and is useful for some special applications [9].

The above methods for temperature sensing were based on the thermal expansion of silica optical fiber. There was another kind of method, which determined temperature by measuring the thermal-induced changes of materials inside the FP cavity.

By changing the gas pressure within the FP cavity, ambient temperature can be determined by measuring the slope of the spectral shift versus pressure. This method, in principle, is not sensitive to the changes of the cavity length. Thus, strain-insensitive temperature sensing of up to 1000°C was performed with strain of up to 3600 με [10]. Similarly, Liu et al. [11] developed a fiber-optic temperature sensor based on differential pressure detection by using a silicon diaphragm. The temperature-related pressure changes of the sealed air introduced the deformation of the silicon diaphragm, which can be measured by the wavelength shift of the FFPI sensor.

Liquid, instead of air, can also be filled into the FFPI cavity, and its thermo-optical characteristics can be used for temperature sensing. Zhao et al. [12] developed an FFPI temperature sensor by filling anhydrous ethanol in the FP cavity. The refractive index of ethanol, thus the phase shift of the reflected beam from the FFPI, varied with temperature changes. Good linearity of temperature sensing was demonstrated. Recently, Yang et al. [13] developed an FFPI formed in a mercury-filled silica tube. The mercury-filled tube was fusion spliced with an SMF. The mercury surface and the end face of SMF formed an FFPI. The cavity length decreased when temperature increased and the volume of mercury was expanded. An ultrahigh temperature sensitivity of up to −41.9 nm/°C was achieved.

4.1.2 Strain, Displacement, and Force Sensors

In addition to the sensing parameter of temperature, strain is another universally used parameter that needs to be tested in a lot of applications. Strain is defined as the relative deformation of the element under test, that is, $\Delta L/L$, often with a unit of microstrain (με). FBG strain sensors have been used more often than FFPI sensors, thanks to their high capacity of multiplexing. However, FBGs often have a length of around 10 mm, while FFPI strain sensors can be fabricated with a length of several tens of micrometers. Conventional FBGs cannot withstand temperature higher than 300°C. Therefore, FFPI strain sensors are excellent for some specific applications, especially for those requiring very compact size and operating in high-temperature environments.

Rao and Ran [14] developed FFPI strain sensors for high-temperature application by using laser micromachining technologies, including fs-laser micromachining [15], 157-nm laser micromachining [16], and also PCFs [17]. Strain can be measured with good linearity at a temperature of 800°C, as shown in Figure 4.3.

There were also numerous publications focused on how to enhance the strain sensitivity of FFPI sensors. By splicing a section of hollow-core ring PCF in between two SMFs, an FFPI sensor with strain sensitivity of 15.4 pm/με was achieved [18]. Several other groups, including Yiping Wang [19,20], Benli Yu [21], and their coworkers, developed sensitive FFPI strain sensors based on microbubbles whose dimension and shape were precisely controlled by optimizing the fusion splicing process. The strain sensitivity of FFPI sensors was recently enhanced up to 43 pm/με.

For displacement measurement, a simple scheme was developed by Wang et al. [22] by using SMF as both the input and output fiber. A reflective film, attached to the object under test, together with the end face of the SMF formed an FFPI. The displacement of the object corresponded to the FFPI cavity length and was measured

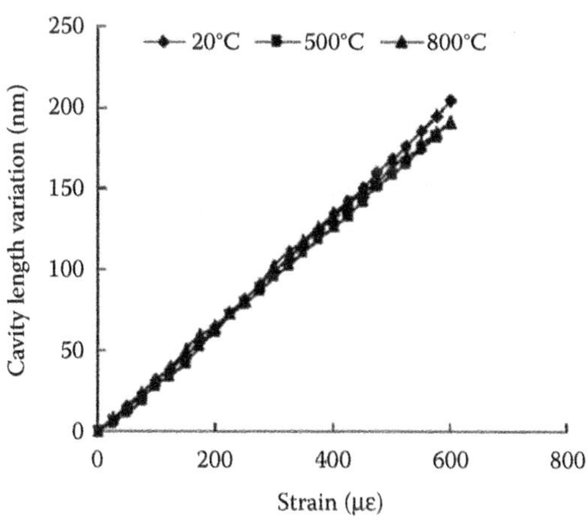

Figure 4.3 High-temperature strain sensing by PCF–FFPI fabricated by 157-nm laser micromachining technology.

by the interferometer interrogating method, with accuracies of 0.05 nm over the dynamic range of 0.005–79.1 nm and 0.5 nm over the dynamic range of 79.1–3200 nm. Similar methods were further investigated by Yu [23], Huang [24], and their coworkers. FFPI sensors can also be used for angle sensing after certain kinds of packaging and mounting [25]. Vibration can be considered as dynamic strain or displacement, which can be measured by integrated cantilever or microstructured beam [26].

Force sensors are of great importance for industrial measurement and also for some bio-applications like micro-surgery. Liu et al. [27] integrated a micro FFPI at the surgery tool tip and the FFPI was interrogated by an optical coherence tomography (OCT) system. The outer diameter of the force sensing tool was less than 1 mm. The calibration characteristics were studied and the dynamic range of force sensing was at the scale of several millinewtons. Recently, Gong et al. [28] developed a sensitive force sensor based on an FFPI with a short section of optical microfiber in the cavity. A short FBG was inscribed and then drawn into optical microfiber with a diameter of several micrometers by the heating-and-drawing method. The far end of the fiber was cleaved so that the broadband Fresnel reflection and the narrowband reflection from the FBG interfere. The use of the optical microfiber enabled an ultrasensitive FFPI force sensor, with a sensitivity of 0.221 pm/μN. Good repeatability of force sensing was demonstrated, and the sensitivity can be controlled by the diameter of the optical microfiber, as shown in Figure 4.4.

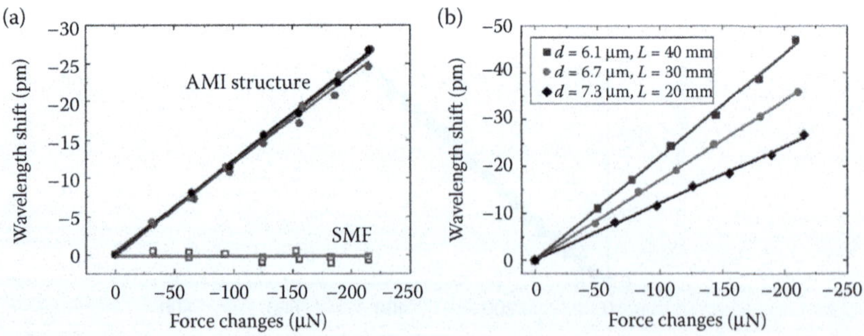

Figure 4.4 Sensitive force sensing based on FFPI with a section of optical microfiber. (a) The sensitivity is much higher than that using a section of single mode fiber instead of the microfiber. (b) Tuning the sensitivity by controlling the diameter of the microfiber.

4.1.3 Pressure Sensors

Pressure sensors have been extensively investigated in the last decades, especially for oil/gas and space applications. Kao and Taylor [29] developed a simple intrinsic FFPI pressure sensor by fixing the FFPI in a cylindrical housing by epoxy, as shown in Figure 4.5. The FFPI was always under tension during the test. The deformation of the diaphragm induced by external air pressure changes was transferred into the strain on the FFPI and was further measured by detecting the phase changes.

Liquid level sensor, which measures pressure at several kPa, is useful for oil storage and many other applications. Lü and Yang [30] developed a liquid level sensor by packaging FFPI in a specially designed mechanical structure, with a resolution of 2 mm over the measurement range of 2 m. The resolution was further enhanced to 0.4 mm [31]. The size of the sensor was large. By using a fused silica ferrule and CO_2 laser heating bonding technology, a simple extrinsic FFPI was formed by aligning the optical fiber with a diaphragm attached to the end of the ferrule [32]. In this case, there was no strain on the FFPI sensor and the deformation of the diaphragm directly corresponded to the phase changes of the FFPI. Similar to other extrinsic FFPIs, it has the advantage of low-temperature cross sensitivity (0.013 nm/°C) with a pressure sensitivity of 5.3 nm/kPa. Due to its high sensitivity, it can be used for liquid level sensing, with a resolution of 0.7 mm over 5-m measurement range [33]. Moreover, the size of the sensor was reduced to 0.5 mm, which can be integrated to other systems easily. Some other structures of the FFPI pressure sensors were designed for special use; for example, 45° angled fiber and cross-axial configuration were used for better surface mounting [34], as shown in Figure 4.6. A packaging design of temperature-insensitive

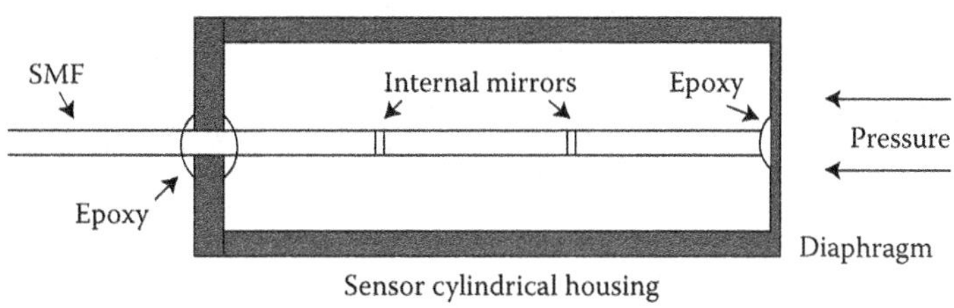

Figure 4.5 Packaged FFPI pressure sensor.

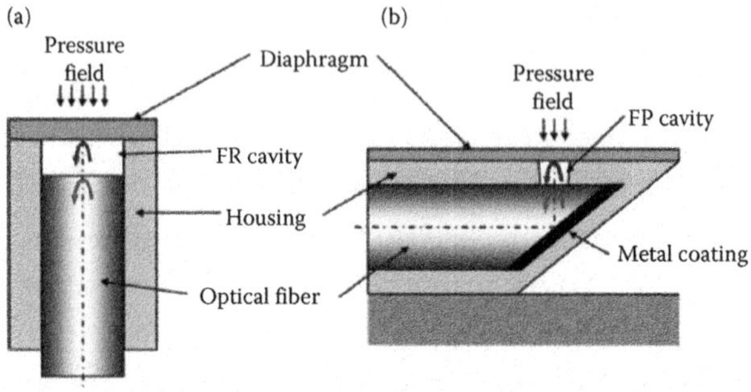

Figure 4.6 FFPI pressure sensor based on (a) coaxial and (b) cross-axial configuration.

FFPI pressure sensor was developed for downhole applications [35]. Pressure of up to 50–60 MPa was measured with a low-temperature sensitivity.

The size of the FFPI pressure sensor can be further reduced by micromachining technologies. Thanks to the compact size, the sensing performance including stability of the sensor can also be enhanced. Different kinds of micromachining technologies for fabricating the FFPIs were introduced in Chapter 2. Here, we discuss the sensing performance for FFPI by different fabrication methods.

Chemical etching has the advantages of low cost and capability for mass production. It is based on the different etching rate of the fiber core and cladding due to their different material composition. An FFPI can be formed by fusion splicing the etched fiber with another cleaved fiber [36]. By cleaving one end of the FFPI and further polishing the fiber end into a thin diaphragm, the FFPI pressure sensor was formed. The pressure sensitivity can be adjusted by the thickness of the diaphragm. Usually, the roughness of the etched bottom of the air cavity is not good enough for high-performance FFPI and needs to be improved by optimizing the charge intensity and duration during the splicing.

By using laser micromachining technologies, the FFPI can be fabricated through similar steps as described above by using the etching method. By using fs-laser micromachining, the fabricated FFPI sensor can work under high temperatures of up to 700°C [37], which is one of the advantages of the micromachined FFPI sensor. Figures 4.7 shows the pressure and temperature sensing performance of it. During the fabrication process, the laser spot needs to be precisely aligned with the optical fiber end, which makes this technology

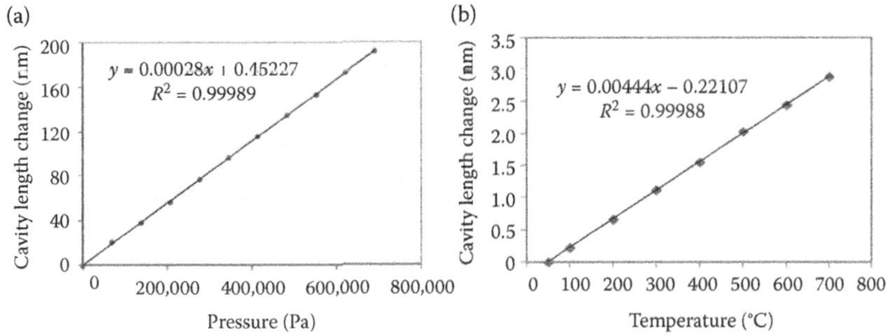

Figure 4.7 (a) Pressure and (b) temperature sensing performance of the laser micromachined FFPI sensor.

complex and relatively high cost. Therefore, it is often used for harsh environment sensing with high-performance requirements.

There were also other kinds of FFPI pressure-sensing structures fabricated by simply cleaving and splicing. One example is by splicing a section of silica rod between two SMFs [38]. The silica rod acted only as a support for forming the FFPI structure. However, this structure has relatively high cross talk between pressure and temperature because both parameters had a strong effect on the readout signal, that is, the spectral shift of the FFPI. Another hybrid optical fiber FFPI structure was fabricated, and a passive temperature compensation scheme was used to reduce 85% of the temperature dependence [39]. By detecting the length of the air gap, a pressure sensitivity of 0.316 nm/psi was obtained. The temperature cross sensitivity was about 0.026 psi/°C.

There are two schemes of FFPI for transferring the pressure variation to the air cavity length changes of the FFPI. One is by using a thin diaphragm at the end of a hollow or etched optical fiber, introduced as above. The pressure is added on the diaphragm along the axis of the fiber, and the deformation of the thin diaphragm corresponds to the cavity length changes. The other is by adding pressure on the lateral direction of the sensing structure. Early trials were based on aligning two optical fiber ends in a silica capillary or ferrule. Pressure was exerted on the lateral direction of the capillary and induced axial strain of the FFPI so that it changed the cavity length of the FFPI.

In order to make the sensing structure compact and reliable, PCFs or other kinds of microstructured fibers were used instead of hollow silica ferrules. PCFs can be fusion spliced in between two SMFs and form an FFPI [17]. Pressure on the lateral direction can be measured by detecting the wavelength shift induced by the axial strain. Frazao [40], Guan [41], and their coworkers demonstrated FFPI pressure sensors using different kinds of PCFs, and sensitivities of 0.82 nm/MPa and 0.023 nm/MPa were obtained, respectively.

Different kinds of materials were used for fabricating the FFPI pressure sensors. Polymers have advantages of being easy-to-fabricate, and are also easily deformed so that FFPI sensors based on polymer materials may have high sensitivity of pressure. Eom et al. [42] developed an FFPI pressure sensor by packaging a lensed fiber and a hybrid polymer thin diaphragm in a glass ferrule. The cavity length changes were detected to determine the external pressure. Highly deformable material, PDMS, was used to improve the sensitivity. A high sensitivity of 1.41 μm/kPa and a detection resolution of 0.03 kPa were obtained.

Wang et al. [43] developed an FFPI pressure sensor for underwater blast wave pressure detection, by coating at the end face of an optical fiber with two reflective films sandwiched with a polymer spacer with a thickness of 50 μm. The repeatability is 1.82% within the full range of 0–55 MPa. The response time is 0.767 μs. Bae and Yu [44] developed a UV-molding method for fabricating the FFPI pressure sensor. The fabrication process is simple and the calibration curve was achieved with excellent linearity.

Besides the polymer materials, silicon has also been used frequently to achieve unique properties of sensing. Liu and Han [45] developed an FFPI gas pressure sensor by attaching a silicon pillar to the optical fiber tip. The pressure was determined by measuring the temperature of the silicon, which was heated by a visible laser beam. In this experiment, the unique absorption property of silica was used, that is, high absorption at visible wavelength for heating and low absorption at near infrared band for generating two-beam interference.

Other kinds of special FFPI pressure sensors were developed for shock pressure testing with fast response time. More than 10 years ago, Gander et al. [46] fabricated an FFPI pressure sensor by forming the cavity with a cleaved SMF and a 2-μm-thickness copper diaphragm and sealing the structure by epoxy. The frequency response of the sensor on the dynamic pressure was measured. Shock wave was investigated with an FFPI pressure sensor with response time at the millisecond level, which used a silicon dioxide diaphragm as the transducer. Further, by sealing a laser micromachined cavity with aluminized polycarbonate diaphragm, an FFPI pressure sensor with a fast response time of 3 μs was developed [47].

4.1.4 *Acoustics and Ultrasonic Sensors*

The acoustic or ultrasonic parameters can be measured based on aforementioned dynamic FFPI pressure sensors with fast response time. Usually, the packaging of the FFPI acoustic or ultrasonic sensors was also similar to that of the pressure sensors, except the material and the geometric dimensions of the diaphragm [48]. An FFPI acoustic sensor based on a multi-layer graphene diaphragm was developed by Ma et al. [49], as shown in Figure 4.8. Experimental results indicated a pressure-induced deflection of 1100 nm/kPa and a noise equivalent acoustic signal of ~60 μ Pa/Hz$^{1/2}$ at the frequency of 10 kHz. The sensor had a flat frequency response in the range of 0.2–22 kHz. By using a similar structure but with a gold-coated polyethylene terephthalate (PET) diaphragm, an FFPI acoustic sensor was developed, with a sensitivity of 40 mV/Pa at 1 kHz [50]. The frequency response range was 100 Hz to 12.5 kHz and the signal-to-noise ratio was 60 dB. Another polymer material poly(phthalazinone ether sulfone ketone) (PPESK) diaphragm was also used for fabricating the FFPI acoustic sensor, and the sensing performance was investigated [51].

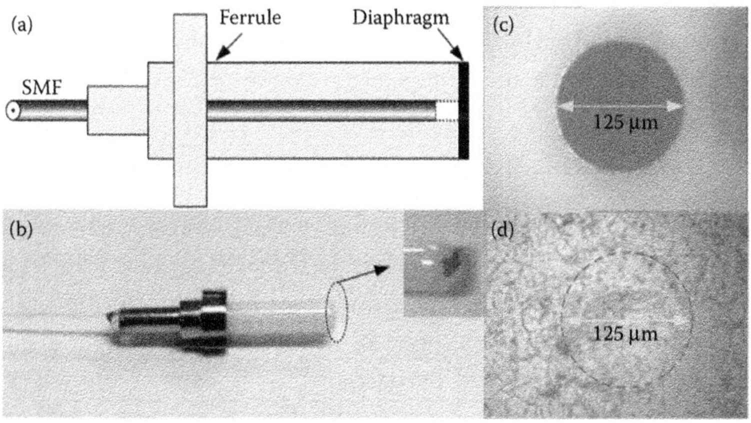

Figure 4.8 (a) Schematic structure and (b) a photo of the FFPI acoustic sensor based on a multi-layer graphene diaphragm. (c) The endface of the ferrule and (d) the multi-layer graphene film.

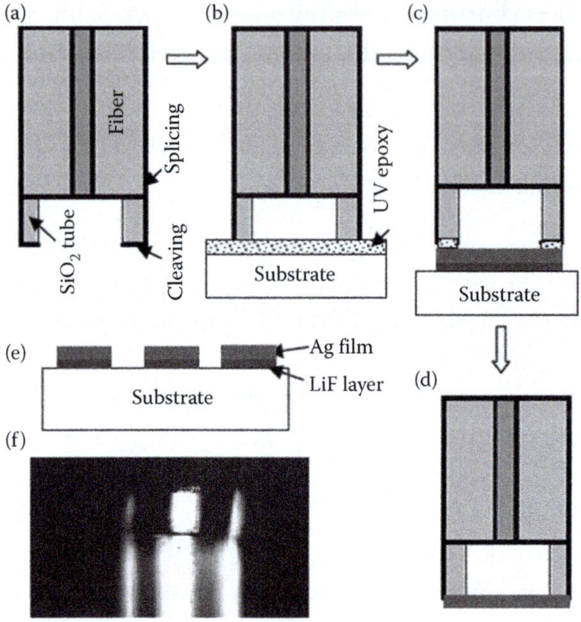

Figure 4.9 Fabrication process of FFPI acoustic sensor. (a) Fusion splicing and cleaving, (b)–(d) the silver film attached to the silica tube by UV glue, (e) the silver film deposited on the silica substrate, and (f) an optical microscopic picture of the fabricated sensor.

The fabrication process and the structure of an FFPI ultrasonic sensor are shown in Figure 4.9. The 300-nm-thick diaphragm was fabricated by thermal deposition method and then transferred onto the fiber tip with a hollow silica tube [52]. By detecting the wavelength shift, a static pressure sensitivity of 1.6 nm/kPa and a resonant frequency of 1.44 MHz were obtained. Another earlier experiment demonstrated an acoustic noise floor of 2.3 kPa and a frequency response of 25 MHz [53].

4.1.5 Electric- and Magnetic-Related Sensors

Electric-related parameters, including partial charges, current, and electric field, are of great importance in the field of energy and power transmission. Partial charge detection in oil-filled transformers can be performed by the aforementioned FFPI acoustic sensors. The acoustic wave bandwidth generated by partial discharges is at the 100 kHz level. Song et al. [54] developed an FFPI acoustic sensor for partial charge detection, by securing SMF in a silica tube and then sealing the tube with a deformable silica diaphragm.

Magnetic fluid can be used as a sensitive medium for current detection. Xia et al. [55] demonstrated an FFPI current sensor by measuring the magnetic-induced refractive index changes of the magnetic fluid. A detection resolution of 1 mA was obtained.

FFPI electric field sensor was developed by measuring the deflection of the micro cantilever, which also served as one of the reflective surfaces for the FFPI. Priest et al. [56] fabricated an FFPI sensor by aligning SMF and a 17-μm-thickness aluminum cantilever. The displacement of the cantilever was induced by the external electric field and measured by the changes of the FFPI interference fringes.

Except for electric-related parameters, FFPI sensors are also promising for weak magnetic field detection. Anbo Wang and coworkers developed an FFPI sensor by using Metglas ($Fe_{77.5}B_{15}Si_{7.5}$) wire as both the magnetic transducer and one of the reflectors [57]. DC magnetic fields of 100–35,000 nT were measured with low-vibration cross sensitivity and 99% compensation of thermal-induced fluctuations. A resolution of 50 nT was further achieved, and also a related theoretical model was developed for the magnetic field sensing [58].

Zhao et al. [59] developed an FFPI magnetic field sensor by filling the FP cavity with magnetic fluid. The refractive index of the magnetic fluid increases as the magnetic field increases. The measurement resolution of 0.5 Gs over the range of 0–400 Gs was achieved (1 Gs = 10^{-4} T). Further, the temperature compensation was obtained by integrating the FFPI structure with an FBG [60]. Zhang et al. [61] developed a simple FFPI magnetic field sensor, which was capable of measuring magnetic field strength up to 560 Oe with good linearity. A sensitivity of 854 pm/Oe was obtained.

4.2 Biochemical Sensors

Fiber-optic biochemical sensors have been extensively investigated for more than 30 years [62]. Wolfbeis has written several review articles on this topic reflecting the progress in this field [63–69]. Actually, the variety of the biochemical parameters is much more diverse than that of the physical parameters, since numerous kinds of analytes need to be measured by biochemical sensors. The specificity for biochemical sensing requires specific sensitive materials being used for certain kinds of analytes. Considering the combination of different fiber microstructures and sensitive materials, the types of fiber-optic biochemical sensors are rather huge. Here, we focus on the fiber-optic FP biochemical sensors for detecting parameters including refractive index, humidity, and gas concentrations.

4.2.1 Refractive Index Sensors

Refractive index is one of the most important parameters for optical waveguiding and photonic devices and is also significant for biochemical sensing as it is directly related to concentration of gas or other biochemical species. Refractive index can be measured precisely with FFPI sensors. One of the most commonly used FFPI schemes is by filling the gas or liquid into the FP cavity and measuring the phase shift by detecting the interference wavelength shift. There are two ways of filling the sample into the cavity. One is through the holes at the end of the FP cavity, that is, end face sampling, while the other is through the side wall of the cavity, that is, side sampling.

By using the solid-core PCF or other microstructured fibers [70,71], the solid fiber core can provide sufficient reflection for the FFPI interference and the hollow holes around the core can serve as the sampling channels. Tian et al. [71] developed an FFPI by sandwiching silica tube with two microstructured fibers, which had two microfluidic sampling holes in the cladding. One end of the microstructured fiber was cleaved with an angle to reduce the influence of reflection from the third surface. A vacuum pump was used for drawing the liquid into the FP cavity. Good interference fringes were obtained, and a high sensitivity of 1051 nm/RIU was achieved. A similar structure was used for measuring the refractive index of gas at different pressures.

Sampling through the side wall of the FP cavity may be simpler than end face sampling. Duan et al. [72] developed an open-cavity FFPI refractive index sensor by large lateral offset splicing a short section of SMF between two SMFs. The fabrication process is simple and a sensitivity of 1540 nm/RIU was obtained. Wu et al. [73] developed an open-cavity FFPI sensor by splicing a section of C-shaped fiber in between two SMFs and got a refractive index sensitivity of 1368 nm/RIU. Wieduwilt et al. [74] fabricated a micro FFPI cavity on the fiber taper by focused ion beam milling method and then coating the cavity end faces with improved reflectivity. An ultrahigh sensitivity of 11,500 nm/RIU was achieved.

Different from the previous FFPIs with two-beam interference, Ran et al. [75] and Gong et al. [76] developed FFPIs with three-beam interference. The theoretical models were developed for better understanding of the influence of the FFPI structural parameters on the performance of the FFPI sensing based on three-beam interference. One unique characteristic of the three-beam interferometric FFPI sensor is the high fringe contrast. As is well-known, the fringe contrast of the two-beam interference reaches its maximum value when the reflectance of the two surfaces equals, which is a strict constraint. In order to get a high fringe contrast, the requirement on the reflectance of the three surfaces for the three-beam interference can be expressed as [76]

$$-1 \leq \frac{R_{\mathrm{I}} + R_{\mathrm{II}} - R_{\mathrm{III}}}{2\sqrt{R_{\mathrm{I}} R_{\mathrm{II}}}} \leq 1 \tag{4.1}$$

R_{I}, R_{II}, and R_{III} are the reflectances of the three surfaces. Unlike the two-beam interference, the constraint condition for the three-beam interference to obtain the optimal fringe contrast is an inequality. This makes it easier for the three-surface-based hybrid GI-FFP sensor to obtain high performance than the conventional two-surface FFP sensors.

Another advantage of the three-beam interferometric FFPI sensors is its capability for dual-parameter sensing or temperature-insensitive refractive index sensing. It can measure the refractive index by detecting the fringe contrast changes, while measuring temperature via the wavelength shift of the FFPI sensor [77–81].

Quan et al. [82] developed such an FFPI sensor for detecting the refractive index of gas. The structure of the FFPI is shown in Figure 4.10. They fusion spliced a hollow tube and a solid-core PCF to the SMF in sequence. The FP interference occurred between the end faces of the SMF and the cleaved PCF. The use of the hollow tube formed a cavity for holding the gas samples. The refractive index changed with the gas pressure. The experimental results demonstrated a sensitivity of 30,899 nm/RIU.

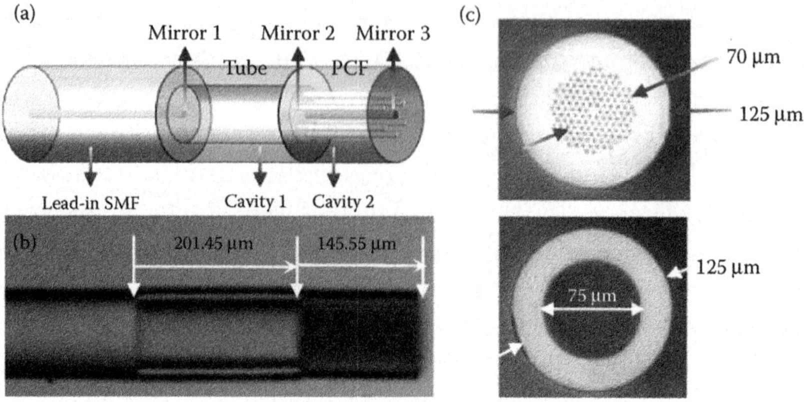

Figure 4.10 FFPI refractive index sensors based on end face sampling. (a) Schematic structure and (b) microscopic picture of the fabricated FFPI sensor. (c) The cross-section of the photonic crystal fiber and the hollow micro tube.

4.2.2 Humidity Sensors

In Chapter 2, we have introduced film-based FFPI structures. Different kinds of sensitive films can be used for the measurement of a variety of parameters. For humidity sensing, cellulose acetate butyrate (CAB) and chitosan were often used as sensitive materials [83]. Nanocomposite polyacrylamide can also be used for humidity sensing.

Xu et al. [84] developed an FFPI humidity sensor by coating the CAB film, with a thickness of about 50 μm, at the end of SMF, as shown in Figure 4.11. The fabrication process was very simple. A relative humidity of 8.8%–88.1% was measured. Long stability was

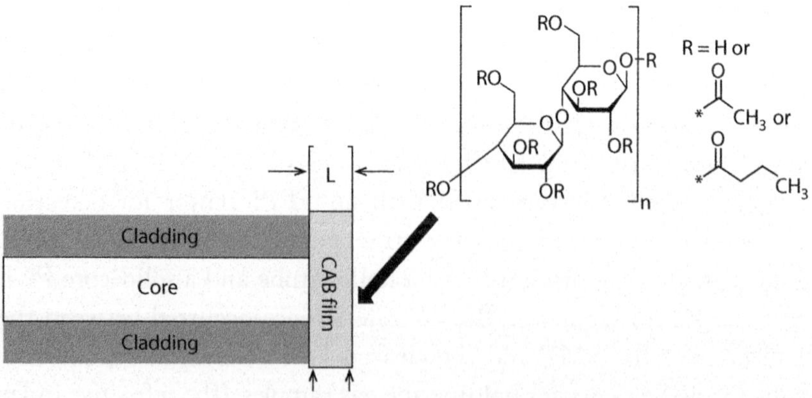

Figure 4.11 FFPI refractive index sensor based on CAB film.

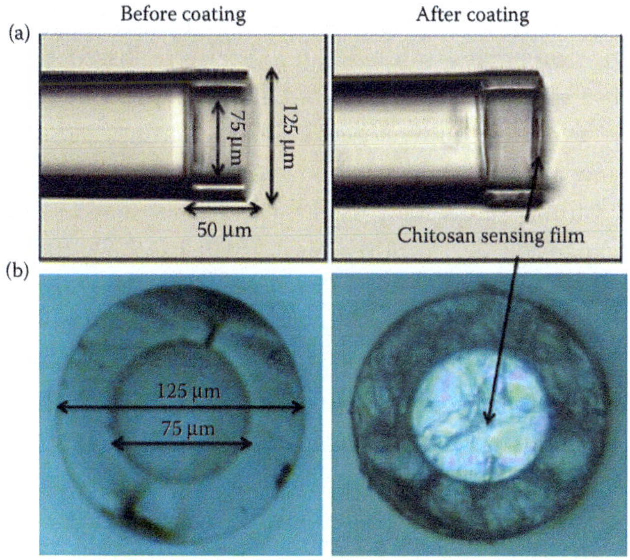

Figure 4.12 Extrinsic FFPI humidity sensor based on chitosan film. The microscopic pictures of (a) the side-view and (b) the cross-section of the FFPI sensor before and after coating.

demonstrated with a precision of 0.03% over 100 min. The response time was around 2.15 s/%RH.

Chen et al. [85] developed an FFPI humidity sensor by using an extrinsic FP cavity, which was formed by fusion splicing a hollow tube with the SMF and then sealing the end of the tube via coating a thin chitosan film, as shown in Figure 4.12. The cavity length was controlled by cleaving the hollow tube under the optical microscope. The thickness of the chitosan film was about 6.2 μm. The refractive index of chitosan is about 1.45 so that the film can provide a Fresnel reflection similar to that of the SMF end, thus getting a good fringe contrast for FFPI sensing. A sensitivity of 0.13 nm/%RH was obtained for relative humidity ranging from 20%RH to 95%RH. Thanks to the thinness of the film, a fast response time of 380 ms was observed.

4.2.3 Gas Sensors

Gas sensors are useful for the chemical industry, environmental monitoring, and even health diagnostics. Similar to humidity sensing, gas sensors also employ films that are sensitive to certain types of gases.

Fiber-optic gas sensors are special for flammable gas detection as they are inherently safe and unlike electric-based sensors, will never cause sparks during detection. Therefore, FFPI hydrogen sensors were extensively investigated for developing and using clean energy, and also for the semiconductor and other industrial fields. Pd-related films were widely used for hydrogen sensing due to their volume expansion (up to 900 times) after absorbing the hydrogen. Yang et al. [86] coated an FFPI structure with Pd–Ag film, which absorbed hydrogen and added mechanical stress onto the FFPI. The tiny cavity length changes can be detected, and the hydrogen concentration was determined. The sensitivity is not high as the radial size of the FFPI was large, hundreds of micrometers, compared to the 100-nm-thickness film.

Yang and coworkers [87,88] contributed a lot to the development of FFPI hydrogen sensors. They tried different kinds of hydrogen-sensitive materials and also various fiber microstructures. Pt/WO_3 film was coated on the capillary, in which an FFPI was formed based on polymer materials. The exposure of Pt/WO_3 film in hydrogen will generate heat, introducing a local temperature rise, which was further measured by the sensitive FFPI sensor via the wavelength shift detection. A high sensitivity of -5.1 nm/% was achieved at the low hydrogen concentration range of 0%–0.5%. Further, the sensitivity was enhanced to –10 nm/% (Figure 4.13), corresponds to 1 pm/ppm and a temperature sensitivity of –5.5 nm/°C, which is very high compared to the intrinsic temperature sensitivity of ~10 pm/°C for

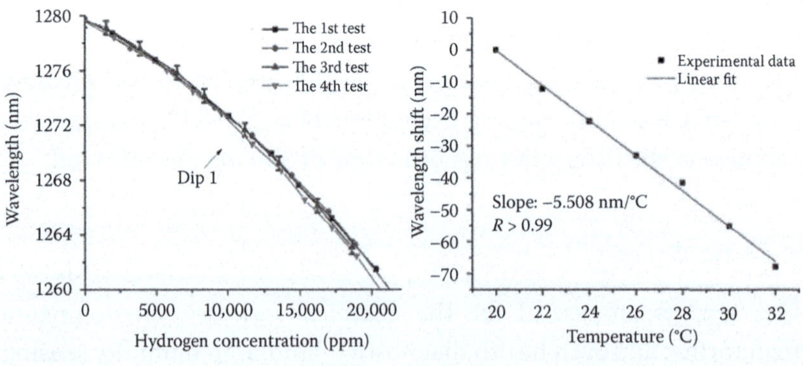

Figure 4.13 Highly sensitive FFPI hydrogen sensor based on Pt/WO_3 film and its temperature responses.

FBGs. Besides the hydrogen detection, there were also FFPI sensors developed for other kinds of gas sensing, including volatile organic compounds (VOCs) and mixed gases [89–92].

4.2.4 Other Types of Biochemical FFPI Sensors

There exists huge demand for the detection of numerous biochemical analytes and related parameters. FFPI sensors have great potential for exploring this field, and a few examples have been demonstrated for surface tension analysis [93], thermo-optic coefficient measurement [94], and dissolved organics detection [95]. Optical label-free methods like refractive index detection are only a small group of biochemical sensing examples. There are several other technologies being used for biochemical detection.

4.3 Temperature-Insensitive or Temperature-Compensated Sensing

Temperature variation exists everywhere for the field trial of optical fiber sensors. Therefore, it is very important to develop temperature-insensitive or temperature-compensated sensors for detecting all kinds of other parameters. Great contributions have been made to solve this problem. There are mainly three strategies. The first is by using extrinsic FFPI sensors, which have low-temperature sensitivity. The second is by designing the packaging or structure of the FFPI so that the temperature effect can be compensated. The third is by measuring the temperature-insensitive readout parameter, like the fringe contrast in dB.

It is known that extrinsic FFPI has low-temperature sensitivity, compared with their intrinsic counterparts. It comprises a hollow cavity with two reflective mirrors, often formed simply by cleaved optical fiber with the Fresnel reflection. The reasons for the low-temperature sensitivity include two aspects. One is the thermally induced changes on the refractive index of air is smaller than that of silica. The other is the inherent compensation effect, that is, as temperature increases, the thermo-expansion of the mirrors of the FP cavity tends to reduce the cavity length, while the expansion of the side wall of the cavity tends to increase the cavity length [97]. By using extrinsic FFPI sensors, the influence of temperature can be neglected when the requirement for measurement accuracy is not very high.

The second strategy is by compensating the temperature effect via designing the sensor structure [35,38,98]. One temperature compensation method is by packaging the FFPI sensor with a specially designed structure. A metal wire was used and its end face served as one of the reflective surfaces for the FFPI [35]. The expansion of the outer tube for alignment tends to make the cavity length longer, while the expansions of the optical fiber and the metal wire tend to make the cavity length shorter. Thus, the temperature effect was compensated. Temperature compensation of FFPI sensors can also be achieved by using hollow-core fiber and different kinds of fibers.

Another design was often used for pressure sensors, as shown in Figure 4.14 (lower picture) [35]. The two ends of the FFPI were attached to the inner tube and middle tube, respectively. As temperature increased, the expansions of the two tubes introduced similar displacement in the same direction onto each end of the FFPI. Therefore, the thermal-induced strain on the FFPI can be compensated.

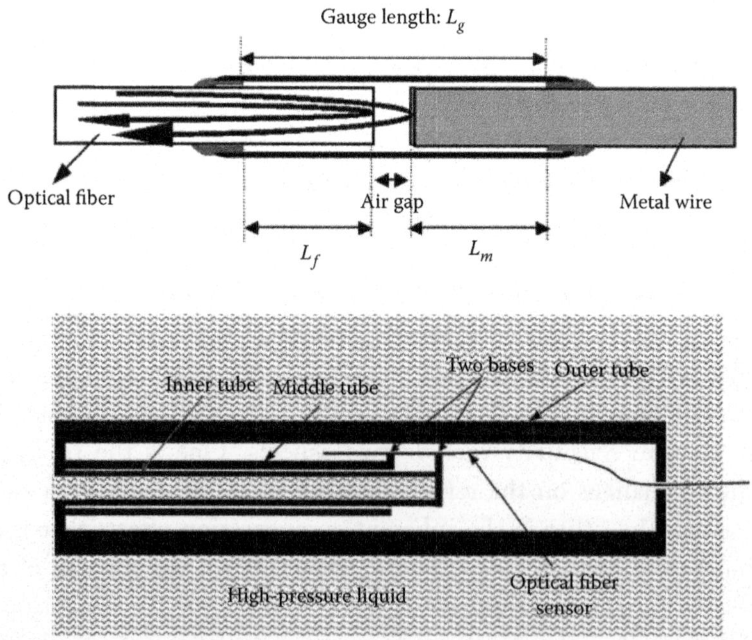

Figure 4.14 Packaging of the FFPI for temperature compensation.

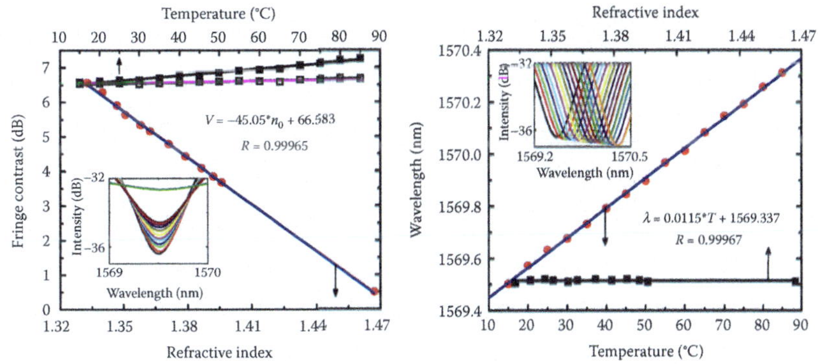

Figure 4.15 Dual-parameter (temperature and refractive index) sensing by FFPI via simultaneous detection of fringe contrast and wavelength shift.

The third strategy is by measuring the fringe contrast of the reflective spectra of the FFPI, which is not influenced by the temperature variations [75,78,99–101]. Thus, the FFPI can perform temperature-insensitive measurement. This method is often used for refractive index sensing. By using the FFPI based on three-beam interference, the refractive index and temperature can be measured by the fringe contrast changes and the wavelength shift of the reflective spectra, respectively, as shown by the results in Figure 4.15 [78]. The cross talk between the two measured parameters is very low.

Besides the refractive index, FFPI sensors can be designed and used for simultaneous measurements of temperature and another parameter like relative humidity, pressure, or strain [102–104]. The main idea is that the dependences, that is, the sensitivity coefficient, of wavelength shift and the fringe contrast on the two parameters are different. Thus, a 2×2 coefficient matrix can be used for data processing [104].

References

1. Lee, C. E., Gibler, W. N., Atkins, R. A., Alcoz, J. J., and Taylor, H. F. 1991. Metal-embedded fiber-optic Fabry–Perot sensors. *Optics Letters*, 16(24), 1990–1992.
2. Lee, C. E., Markus, A. M., Udd, E., and Taylor, H. F. 1989. Optical-fiber Fabry–Perot embedded sensor. *Optics Letters*, 14(21), 1225–1227.
3. Zhang, J., Sun, H., Rong, Q., Ma, Y., Liang, L., Xu, Q., Zhao, P., Feng, Z. Y., Hu, M. L., and Qiao, X. 2012. High-temperature sensor using a Fabry–Perot interferometer based on solid-core photonic crystal fiber. *Chinese Optics Letters*, 10(7), 070607.

4. Ding, W., Jiang, Y., Gao, R., and Liu, Y. 2015. High-temperature fiber-optic Fabry–Perot interferometric sensors. *Review of Scientific Instruments*, 86(5), 055001.
5. Zhu, T., Ke, T., Rao, Y., and Chiang, K. S. 2010. Fabry–Perot optical fiber tip sensor for high temperature measurement. *Optics Communications*, 283(19), 3683–3685.
6. Tian, Z., Yu, Z., Liu, B., and Wang, A. 2016. Sourceless optical fiber high temperature sensor. *Optics Letters*, 41.2, 195–198.
7. Lee, D., Yang, M., Huang, C., and Dai, J. 2014. Optical fiber high-temperature sensor based on dielectric films extrinsic Fabry–Perot cavity. *Photonics Technology Letters, IEEE*, 26(21), 2107–2110.
8. Li, L., Feng, Z., Qiao, X., Yang, H., Wang, R., Su, D., Wang, Y. et al. 2015. Ultrahigh sensitive temperature sensor based on Fabry–Pérot interference assisted by a graphene diaphragm. *IEEE Sensors Journal*, 15(1), 505–509.
9. Liu, G., Han, M., and Hou, W. 2015. High-resolution and fast-response fiber-optic temperature sensor using silicon Fabry–Pérot cavity. *Optics Express*, 23(6), 7237–7247.
10. Lu, Y., Han, M., and Tian, J. 2014. Fiber-optic temperature sensor using a Fabry–Pérot cavity filled with gas of variable pressure. *IEEE Photonics Technology Letters*, 26(8), 757–760.
11. Liu, T., Yin, J., Jiang, J., Liu, K., Wang, S., and Zou, S. 2015. Differential-pressure-based fiber-optic temperature sensor using Fabry–Perot interferometry. *Optics Letters*, 40(6), 1049–1052.
12. Zhao, Y., Wang, D., and Lv, R. 2013. A novel optical fiber temperature sensor based on Fabry–Perot cavity. *Microwave and Optical Technology Letters*, 55(10), 2487–2490.
13. Yang, K., He, J., Wang, Y., Liu, S., Liao, C., Li, Z., Yin, G., Sun, B., and Wang, Y. 2015. Ultrasensitive temperature sensor based on a fiber Fabry–Pérot interferometer created in a mercury-filled silica tube. *IEEE Photonics Journal*, 7.6, 1–9.
14. Rao, Y. J. and Ran, Z. L. 2011. Fibers for sensing-fiber-optic Fabry–Perot sensors take the strain under high temperature. *Laser Focus World*, 47(11), 71.
15. Rao, Y. J., Deng, M., Duan, D. W., Yang, X. C., Zhu, T., and Cheng, G. H. 2007. Micro Fabry–Perot interferometers in silica fibers machined by femtosecond laser. *Optics Express*, 15(21), 14123–14128.
16. Ran, Z. L., Rao, Y. J., Deng, H. Y., and Liao, X. 2007. Miniature in-line photonic crystal fiber etalon fabricated by 157 nm laser micromachining. *Optics Letters*, 32(21), 3071–3073.
17. Rao, Y. J., Zhu, T., Yang, X. C., and Duan, D. W. 2007. In-line fiber-optic etalon formed by hollow-core photonic crystal fiber. *Optics Letters*, 32(18), 2662–2664.
18. Ferreira, M. S., Bierlich, J., Kobelke, J., Schuster, K., Santos, J. L., and Frazão, O. 2012. Towards the control of highly sensitive Fabry–Pérot strain sensor based on hollow-core ring photonic crystal fiber. *Optics Express*, 20(20), 21946–21952.

19. Liu, S., Yang, K., Wang, Y., Qu, J., Liao, C., He, J., Li, Z. et al. 2015. High-sensitivity strain sensor based on in-fiber rectangular air bubble. *Scientific Reports*, 5, 7624.
20. Liu, S., Wang, Y., Liao, C., Wang, G., Li, Z., Wang, Q., Zhou, J. et al. 2014. High-sensitivity strain sensor based on in-fiber improved Fabry–Perot interferometer. *Optics Letters*, 39(7), 2121–2124.
21. Yin, C., Cao, Z., Zhang, Z., Shui, T., Wang, R., Wang, J., Lu, L., Zhen, S. L., and Yu, B. 2014. Temperature-independent ultrasensitive Fabry–Perot all-fiber strain sensor based on a bubble-expanded microcavity. *Photonics Journal, IEEE*, 6(4), 1–9.
22. Wang, T., Zheng, S., and Yang, Z. 1998. A high precision displacement sensor using a low-finesse fiber-optic Fabry–Pérot interferometer. *Sensors and Actuators A: Physical*, 69(2), 134–138.
23. Zhou, X. and Yu, Q. 2011. Wide-range displacement sensor based on fiber-optic Fabry–Perot interferometer for subnanometer measurement. *Sensors Journal, IEEE*, 11(7), 1602–1606.
24. Chen, J. H., Huang, X. G., Zhao, J. R., Tao, J., He, W. X., and Liu, S. H. 2010. Fabry–Perot interference-based fiber-optic sensor for small displacement measurement. *Optics Communications*, 283(17), 3315–3319.
25. Lű, T., Li, Z., Du, Q., and Bi, J. 2008. Fiber-optic angle sensor based on an extrinsic Fabry–Perot cavity. *Sensors and Actuators A: Physical*, 148(1), 83–87.
26. Zhang, Q., Zhu, T., Hou, Y., and Chiang, K. S. 2013. All-fiber vibration sensor based on a Fabry–Perot interferometer and a microstructure beam. *JOSA B*, 30(5), 1211–1215.
27. Liu, X., Iordachita, I. I., He, X., Taylor, R. H., and Kang, J. U. 2012. Miniature fiber-optic force sensor based on low-coherence Fabry–Pérot interferometry for vitreoretinal microsurgery. *Biomedical Optics Express*, 3(5), 1062–1076.
28. Gong, Y., Yu, C. B., Wang, T. T., Liu, X. P., Wu, Y., Rao, Y. J., Zhang, M. L., Wu, H. J., Chen, X. X., and Peng, G. D. 2014. Highly sensitive force sensor based on optical microfiber asymmetrical Fabry–Perot interferometer. *Optics Express*, 22(3), 3578–3584.
29. Kao, T. W. and Taylor, H. F. 1996. High-sensitivity intrinsic fiber-optic Fabry–Perot pressure sensor. *Optics Letters*, 21(8), 615–617.
30. Lü, T. and Yang, S. 2007. Extrinsic Fabry–Perot cavity optical fiber liquid-level sensor. *Applied Optics*, 46(18), 3682–3687.
31. Lü, T., Li, Z., Xia, D., He, K., and Zhang, G. 2009. Asymmetric Fabry–Pérot fiber-optic pressure sensor for liquid-level measurement. *Review of Scientific Instruments*, 80(3), 033104.
32. Lai, C. W., Lo, Y. L., Yur, J. P., and Chuang, C. H. 2012. Application of fiber Bragg grating level sensor and Fabry–Perot pressure sensor to simultaneous measurement of liquid level and specific gravity. *IEEE Sensors Journal*, 12(4), 827–831.
33. Wang, W. and Li, F. 2014. Large-range liquid level sensor based on an optical fibre extrinsic Fabry–Perot interferometer. *Optics and Lasers in Engineering*, 52, 201–205.

34. Bae, H., Zhang, X. M., Liu, H., and Yu, M. 2010. Miniature surface-mountable Fabry–Perot pressure sensor constructed with a 45 angled fiber. *Optics Letters*, 35(10), 1701–1703.
35. Aref, S. H., Latifi, H., Zibaii, M. I., and Afshari, M. 2007. Fiber optic Fabry–Perot pressure sensor with low sensitivity to temperature changes for downhole application. *Optics Communications*, 269(2), 322–330.
36. Pevec, S. and Donlagic, D. 2012. Miniature all-fiber Fabry–Perot sensor for simultaneous measurement of pressure and temperature. *Applied Optics*, 51(19), 4536–4541.
37. Zhang, Y., Yuan, L., Lan, X., Kaur, A., Huang, J., and Xiao, H. 2013. High-temperature fiber-optic Fabry–Perot interferometric pressure sensor fabricated by femtosecond laser. *Optics Letters*, 38(22), 4609–4612.
38. Xu, J., Wang, X., Cooper, K. L., Gary, G. R., and Wang, A. 2006. Miniature temperature-insensitive Fabry–Perot fiber-optic pressure sensor. *IEEE Photonics Technolology Letters*, 18(9–12), 1134–1136.
39. Silva, S., Coelho, L., and Frazão, O. 2014. An all-fiber Fabry–Pérot interferometer for pressure sensing in different gaseous environments. *Measurement*, 47, 418–421.
40. Ferreira, M. S., Bierlich, J., Lehmann, H., Schuster, K., Kobelke, J., Santos, J. L., and Frazão, O. 2012. Fabry–Pérot cavity based on hollow-core ring photonic crystal fiber for pressure sensing. *Photonics Technology Letters, IEEE*, 24(23), 2122–2124.
41. Jin, L., Guan, B. O., and Wei, H. 2013. Sensitivity characteristics of Fabry–Perot pressure sensors based on hollow-core microstructured fibers. *Journal of Lightwave Technology*, 31(15), 2526–2532.
42. Eom, J., Park, C. J., Lee, B. H., Lee, J. H., Kwon, I. B., and Chung, E. 2015. Fiber optic Fabry–Perot pressure sensor based on lensed fiber and polymeric diaphragm. *Sensors and Actuators A: Physical*, 225, 25–32.
43. Wang, J., Wang, M., Xu, J., Peng, L., Yang, M., Xia, M., and Jiang, D. 2014. Underwater blast wave pressure sensor based on polymer film fiber Fabry–Perot cavity. *Applied Optics*, 53(28), 6494–6502.
44. Bae, H. and Yu, M. 2012. Miniature Fabry–Perot pressure sensor created by using UV-molding process with an optical fiber based mold. *Optics Express*, 20(13), 14573–14583.
45. Liu, G. and Han, M. 2015. Fiber-optic gas pressure sensing with a laser-heated silicon-based Fabry–Perot interferometer. *Optics Letters*, 40(11), 2461–2464.
46. Gander, M. J., MacPherson, W. N., Barton, J. S., Reuben, R. L., Jones, J. D., Stevens, R., Chana, K. S., Anderson, S. J., and Jones, T. V. 2003. Embedded micromachined fiber-optic Fabry–Perot pressure sensors in aerodynamics applications. *Sensors Journal, IEEE*, 3(1), 102–107.
47. Watson, S., Gander, M. J., MacPherson, W. N., Barton, J. S., Jones, J. D., Klotzbuecher, T., Braune, T., Ott, J., and Schmitz, F. 2006. Laser-machined fibers as Fabry–Perot pressure sensors. *Applied Optics*, 45(22), 5590–5596.

48. Wang, W., Wu, N., Tian, Y., Wang, X., Niezrecki, C., and Chen, J. 2009. Optical pressure/acoustic sensor with precise Fabry–Perot cavity length control using angle polished fiber. *Optics Express*, 17(19), 16613–16618.
49. Ma, J., Xuan, H., Ho, H. L., Jin, W., Yang, Y., and Fan, S. 2013. Fiber-optic Fabry–Perot acoustic sensor with multilayer graphene diaphragm. *IEEE Photonics Technology Letters*, 10(25), 932–935.
50. Wang, Q. and Yu, Q. 2010. Polymer diaphragm based sensitive fiber optic Fabry–Perot acoustic sensor. *Chinese Optics Letters*, 8(3), 266–269.
51. Mao, X., Tian, X., Zhou, X., and Yu, Q. 2015. Characteristics of a fiber-optical Fabry–Perot interferometric acoustic sensor based on an improved phase-generated carrier-demodulation mechanism. *Optical Engineering*, 54(4), 046107.
52. Guo, F., Fink, T., Han, M., Koester, L., Turner, J., and Huang, J. 2012. High-sensitivity, high-frequency extrinsic Fabry–Perot interferometric fiber-tip sensor based on a thin silver diaphragm. *Optics Letters*, 37(9), 1505–1507.
53. Beard, P. C. and Mills, T. N. 1996. Extrinsic optical-fiber ultrasound sensor using a thin polymer film as a low-finesse Fabry–Perot interferometer. *Applied Optics*, 35(4), 663–675.
54. Song, L., Wang, Z., Wang, A., Liu, Y., and Cooper, K. L. 2006. Angular dependence of the frequency response of an extrinsic Fabry–Perot interferometric (EFPI) fiber acoustic sensor for partial discharge detection. *Journal of Lightwave Technology*, 24(9), 3433.
55. Xia, J., Wang, Q., Liu, X., and Luo, H. 2015. Fiber optic Fabry–Perot current sensor integrated with magnetic fluid using a fiber Bragg grating demodulation. *Sensors*, 15(7), 16632–16641.
56. Priest, T. S., Scelsi, G. B., and Woolsey, G. A. 1997. Optical fiber sensor for electric field and electric charge using low-coherence, Fabry–Perot interferometry. *Applied Optics*, 36(19), 4505–4508.
57. Oh, K. D., Ranade, J., Arya, V., Wang, A., and Claus, R. O. 1997. Optical fiber Fabry–Perot interferometric sensor for magnetic field measurement. *IEEE Photonics Technology Letters*, 9(6), 797–799.
58. Oh, K. D., Wang, A., and Claus, R. O. 2004. Fiber-optic extrinsic Fabry–Perot dc magnetic field sensor. *Optics Letters*, 29(18), 2115–2117.
59. Zhao, Y., Lv, R. Q., Wang, D., and Wang, Q. 2014. Fiber optic Fabry–Perot magnetic field sensor with temperature compensation using a fiber Bragg grating. *IEEE Transactions on Instrumentation and Measurement*, 63(9), 2210–2214.
60. Lv, R. Q., Zhao, Y., Wang, D., and Wang, Q. 2014. Magnetic fluid-filled optical fiber Fabry–Pérot sensor for magnetic field measurement. *IEEE Photonics Technology Letters*, 26(3), 217–219.
61. Zhang, P., Tang, M., Gao, F., Zhu, B., Fu, S., Ouyang, J., Zhao, Z. et al. 2015. An ultra-sensitive magnetic field sensor based on extrinsic fiber-optic Fabry–Perot interferometer and Terfenol-D. *Journal of Lightwave Technology*, 33(15), 3332–3337.

62. Vurek, G. G., Feustel, P. J., and Severinghaus, J. W. 1983. A fiber optic pCO2 sensor. *Annals of Biomedical Engineering*, 11(6), 499–510.
63. Posch, H. E., Wolfbeis, O. S., and Pusterhofer, J. 1988. Optical and fibre-optic sensors for vapours of polar solvents. *Talanta*, 35(2), 89–94.
64. Wang, X. D. and Wolfbeis, O. S. 2012. Fiber-optic chemical sensors and biosensors (2008–2012). *Analytical Chemistry*, 85(2), 487–508.
65. Wolfbeis, O. S. 2008. Fiber-optic chemical sensors and biosensors. *Analytical Chemistry*, 80(12), 4269–4283.
66. Wolfbeis, O. S. 2006. Fiber-optic chemical sensors and biosensors. *Analytical Chemistry*, 78(12), 3859–3874.
67. Wolfbeis, O. S. 2004. Fiber-optic chemical sensors and biosensors. *Analytical Chemistry*, 76(12), 3269–3284.
68. Wolfbeis, O. S. 2002. Fiber-optic chemical sensors and biosensors. *Analytical Chemistry*, 74(12), 2663–2678.
69. Wolfbeis, O. S. 2000. Fiber-optic chemical sensors and biosensors. *Analytical Chemistry*, 72(12), 81–90.
70. Wang, R. and Qiao, X. 2015. Gas refractometer based on optical fiber extrinsic Fabry–Perot interferometer with open cavity. *IEEE Photonics Technology Letters*, 27(3), 245–248.
71. Tian, J., Lu, Y., Zhang, Q., and Han, M. 2013. Microfluidic refractive index sensor based on an all-silica in-line Fabry–Perot interferometer fabricated with microstructured fibers. *Optics Express*, 21(5), 6633–6639.
72. Duan, D. W., Rao, Y. J., and Zhu, T. 2012. High sensitivity gas refractometer based on all-fiber open-cavity Fabry–Perot interferometer formed by large lateral offset splicing. *JOSA B*, 29(5), 912–915.
73. Wu, C., Liu, Z., Zhang, A. P., Guan, B. O., and Tam, H. Y. 2014. In-line open-cavity Fabry–Pérot interferometer formed by C-shaped fiber for temperature-insensitive refractive index sensing. *Optics Express*, 22(18), 21757–21766.
74. Wieduwilt, T., Dellith, J., Talkenberg, F., Bartelt, H., and Schmidt, M. A. 2014. Reflectivity enhanced refractive index sensor based on a fiber-integrated Fabry–Perot microresonator. *Optics Express*, 22(21), 25333–25346.
75. Ran, Z. L., Rao, Y. J., Liu, W. J., Liao, X., and Chiang, K. S. 2008. Laser-micromachined Fabry–Perot optical fiber tip sensor for high-resolution temperature-independent measurement of refractive index. *Optics Express*, 16(3), 2252–2263.
76. Gong, Y., Zhao, T., Rao, Y. J., Wu, Y., and Guo, Y. 2010. A ray-transfer-matrix model for hybrid fiber Fabry–Perot sensor based on graded-index multimode fiber. *Optics Express*, 18(15), 15844–15852.
77. Rao, Y. J., Deng, M., Duan, D. W., and Zhu, T. 2008. In-line fiber Fabry–Perot refractive-index tip sensor based on endlessly photonic crystal fiber. *Sensors and Actuators A: Physical*, 148(1), 33–38.
78. Gong, Y., Guo, Y., Rao, Y. J., Zhao, T., and Wu, Y. 2010. Fiber-optic Fabry–Perot sensor based on periodic focusing effect of graded-index multimode fibers. *Photonics Technology Letters, IEEE*, 22(23), 1708–1710.

79. Choi, H. Y., Mudhana, G., Park, K. S., Paek, U. C., and Lee, B. H. 2010. Cross-talk free and ultra-compact fiber optic sensor for simultaneous measurement of temperature and refractive index. *Optics Express*, 18(1), 141–149.
80. Jiang, M., Li, Q. S., Wang, J. N., Yao, W. G., Jin, Z., Sui, Q., Shi, J., Zhang, F., Jia, L., and Dong, W. F. 2013. Optical response of fiber-optic Fabry–Perot refractive-index tip sensor coated with polyelectrolyte multilayer ultra-thin films. *Journal of Lightwave Technology*, 31(14), 2321–2326.
81. Jiang, M., Li, Q. S., Wang, J. N., Jin, Z., Sui, Q., Ma, Y., Shi, J. et al. 2013. TiO_2 nanoparticle thin film-coated optical fiber Fabry–Perot sensor. *Optics Express*, 21(3), 3083–3090.
82. Quan, M., Tian, J., and Yao, Y. 2015. Ultra-high sensitivity Fabry–Perot interferometer gas refractive index fiber sensor based on photonic crystal fiber and Vernier effect. *Optics Letters*, 40(21), 4891–4894.
83. Yao, J., Zhu, T., Duan, D. W., and Deng, M. 2012. Nanocomposite polyacrylamide based open cavity fiber Fabry–Perot humidity sensor. *Applied Optics*, 51(31), 7643–7647.
84. Xu, W., Huang, W. B., Huang, X. G., and Yu, C. Y. 2013. A simple fiber-optic humidity sensor based on extrinsic Fabry–Perot cavity constructed by cellulose acetate butyrate film. *Optical Fiber Technology*, 19(6), 583–586.
85. Chen, L. H., Li, T., Chan, C. C., Menon, R., Balamurali, P., Shaillender, M., Neu, B. et al. 2012. Chitosan based fiber-optic Fabry–Perot humidity sensor. *Sensors and Actuators B: Chemical*, 169, 167–172.
86. Yang, Z., Zhang, M., Liao, Y., Tian, Q., Li, Q., Zhang, Y., and Zhuang, Z. 2010. Extrinsic Fabry–Perot interferometric optical fiber hydrogen detection system. *Applied Optics*, 49(15), 2736–2740.
87. Zhang, G., Yang, M., and Wang, Y. 2014. Optical fiber-tip Fabry–Perot interferometer for hydrogen sensing. *Optics Communications*, 329, 34–37.
88. Wang, Y., Yang, M., Zhang, G., Dai, J., Zhang, Y., Zhuang, Z., and Hu, W. 2015. Fiber optic hydrogen sensor based on Fabry–Perot interferometer coated with sol-gel Pt/WO3 coating. *Journal of Lightwave Technology*, 33(12), 2530–2534.
89. Melissinaki, V., Farsari, M., and Pissadakis, S. 2015. A fiber-endface, Fabry–Perot vapor microsensor fabricated by multiphoton polymerization. *IEEE Journal of Selected Topics in Quantum Electronics*, 21(4), 1–10.
90. Elosúa, C., Bariáin, C., Matías, I. R., Arregui, F. J., Luquin, A., Vergara, E., and Laguna, M. 2008. Indicator immobilization on Fabry–Perot nanocavities towards development of fiber optic sensors. *Sensors and Actuators B: Chemical*, 130(1), 158–163.
91. Ding, H., Liang, J., Cui, J., and Wu, X. 2009. A novel fiber Fabry–Perot filter based mixed-gas sensing system. *Sensors and Actuators B: Chemical*, 138(1), 154–159.
92. Liu, J., Sun, Y., and Fan, X. 2009. Highly versatile fiber-based optical Fabry–Perot gas sensor. *Optics Express*, 17(4), 2731–2738.

93. Márquez-Cruz, V. A. and Hernández-Cordero, J. A. 2014. Fiber optic Fabry–Perot sensor for surface tension analysis. *Optics Express*, 22(3), 3028–3038.
94. Lee, C. L., Ho, H. Y., Gu, J. H., Yeh, T. Y., and Tseng, C. H. 2015. Dual hollow core fiber-based Fabry–Perot interferometer for measuring the thermo-optic coefficients of liquids. *Optics Letters*, 40(4), 459–462.
95. Liu, N., Hui, J., Sun, C., Dong, J., Zhang, L., and Xiao, H. 2006. Nanoporous zeolite thin film-based fiber intrinsic Fabry–Perot interferometric sensor for detection of dissolved organics in water. *Sensors*, 6(8), 835–847.
96. Zhang, X. and Peng, W. 2015. Temperature-independent fiber salinity sensor based on Fabry–Perot interference. *Optics Express*, 23(8), 10353–10358.
97. Wang, W., Yu, Q., Li, F., Zhou, X., and Jiang, X. 2012. Temperature-insensitive pressure sensor based on all-fused-silica extrinsic Fabry–Pérot optical fiber interferometer. *IEEE Sensors Journal*, 12(7), 2425–2429.
98. Wang, Z., Jiang, Y., Peng, H., Ma, X., and Cui, L. 2013. A temperature-compensated fibre optic extrinsic Fabry–Perot interferometric displacement sensor for fault measurement in geomechanics. *Measurement Science and Technology*, 24(2), 025104.
99. Jiang, M. S., Sui, Q. M., Jin, Z. W., Zhang, F. Y., and Jia, L. 2014. Temperature-independent optical fiber Fabry–Perot refractive-index sensor based on hollow-core photonic crystal fiber. *Optik—International Journal for Light and Electron Optics*, 125(13), 3295–3298.
100. Ma, Y., Qiao, X., Guo, T., Wang, R., Zhang, J., Weng, Y., Rong, Q. Z., Hu, M. L., and Feng, Z. 2012. Temperature-independent refractive index measurement based on Fabry–Perot fiber tip sensor modulated by Fresnel reflection. *Chinese Optics Letters*, 10(5), 050603.
101. Zhao, J. R., Huang, X. G., He, W. X., and Chen, J. H. 2010. High-resolution and temperature-insensitive fiber optic refractive index sensor based on Fresnel reflection modulated by Fabry–Perot interference. *Journal of Lightwave Technology*, 28(19), 2799–2803.
102. Sun, H., Zhang, X., Yuan, L., Zhou, L., Qiao, X., and Hu, M. 2015. An optical fiber Fabry–Perot interferometer sensor for simultaneous measurement of relative humidity and temperature. *IEEE Sensors Journal*, 15(5), 2891–2897.
103. Kim, K. S., Ismail, Y., and Springer, G. S. 1993. Measurements of strain and temperature with embedded intrinsic Fabry–Perot optical fiber sensors. *Journal of Composite Materials*, 27(17), 1663–1677.
104. Yin, J., Liu, T., Jiang, J., Liu, K., Wang, S., Qin, Z., and Zou, S. 2014. Batch-producible fiber-optic Fabry–Pérot sensor for simultaneous pressure and temperature sensing. *IEEE Photonics Technology Letters*, 26(20), 2070–2073.

5

Interrogation and Multiplexing Techniques for FFP Sensors

5.1 Introduction

Generally, a typical FFP sensing system includes light source, sensor head, guiding optical fiber, sensing signal receiving, and demodulation [1], as shown in Figure 5.1. The interferometric fringe pattern of an FFP sensor is modulated by change of the optical phase difference (OPD) induced by the measurand, such as displacement, temperature, strain, pressure, etc. The crucial issue for interrogating the FFP sensor is to obtain the OPD variation from either the intensity or wavelength shift of the interferometric fringe. According to the way of obtaining the relative OPD variation and absolute OPD, the interrogating methods of FFP sensors could be classified into intensity interrogating, spectral interrogating, interferometric scanning, and white light interferometry (WLI).

5.2 Intensity Interrogating Methods and Instruments

5.2.1 Single-Wavelength Intensity Interrogating Method and Instrument

The schematic diagram of the intensity interrogating method is shown in Figure 5.2, for monitoring the OPD variation of an FFP sensor. Light from a laser diode (LD) or light-emitting diode (LED) passes through a circulator (or a fiber coupler) and is reflected by the FFP. After passing through the circulator again, the reflected light is converted by a photodiode to an electrical signal, which is processed electronically. Specially, it is required that the coherent length of the light source must be less than the OPD of the FFP sensor.

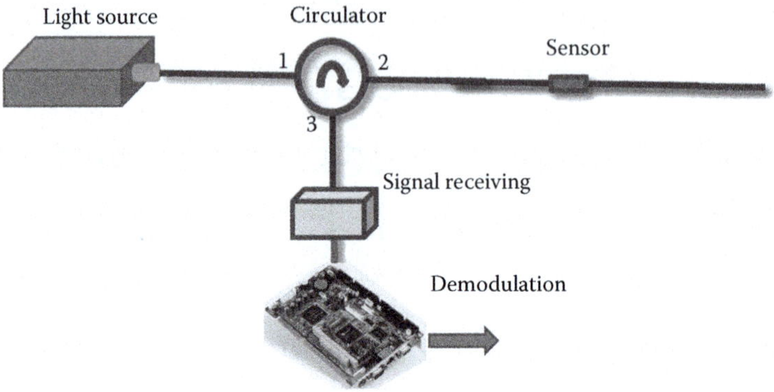

Figure 5.1 Typical configuration of an FFP sensor system.

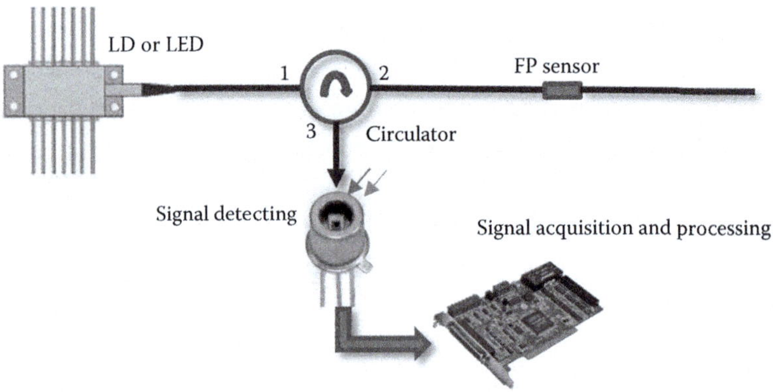

Figure 5.2 Schematic diagram for monitoring the OPD variation of an FFP sensor.

When the incident light intensity is constant, the cavity length of the sensor is a function of the wavelength and the output light intensity. Thus, in the vicinity of point E, the relationship between OPD variation and light intensity reflected is almost linear, as shown in Figure 5.3. Intensity interrogation is quite simple, but it is susceptible to fluctuation of the light intensity, vibration of the optical path, variation of the fiber transmission attenuation, and parameter drift of electrical circuits. An approach based on dual-wavelength carriers to eliminate the drift of the light source and optical/electrical links is proposed [2]. In addition, intensity interrogating puts forward very tight requirements on the consistency of sensors.

In the single-wavelength scheme, the linear dynamic range region of the OPD variation is limited to $<\pi$ rad, as shown in Figure 5.3.

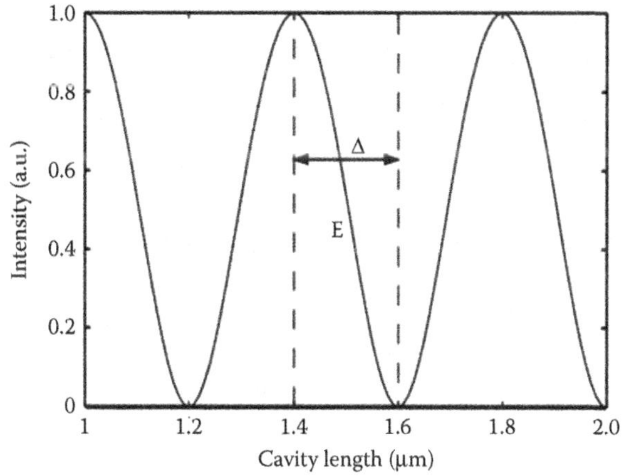

Figure 5.3 Relationship between OPD variation and light intensity reflected.

A passive scheme that makes it possible to overcome this limitation has been implemented using two simultaneously monitored FFP sensors [3]. The interferometers are exposed to the same measurand and use the same light source, but they are fabricated so that their round-trip optical path lengths differ by an odd integral multiple of a quarter wavelength. By suitable processing of the two "quadrature-shifted" optical signals, it is possible to extract information on the measurand-induced phase shift while avoiding sensitivity nulls and ambiguity in the direction of phase change.

5.2.2 Intensity Interrogating Method with Multiple Wavelengths

Multiple wavelengths from one or more cw light sources can be employed as a way of overcoming the sensitivity nulls and direction of change ambiguities that occur when interferometers are interrogated using a single light source. Using a 3-nm wavelength separation between two modes of an AlGaInP laser, Potter et al. [4] achieved quadrature optical outputs from a 20-mm-long FFP cavity. A similar approach using spectrally filtered optical signals from FFP sensors monitored with a low-coherence superluminescent diode (SLD) light source has been reported in which two wavelengths [5] are simultaneously monitored. In another experiment, two wavelengths from an Erbium-doped fiber broadband source are selected by an FBG after reflection from an FFP sensor to provide quadrature optical signals [6]. A three-wavelength-based passive quadrature digital-phase-demodulation scheme is developed for readout of vibration, acoustic, and strain sensors based on FFP [7].

5.3 Spectral Interrogating for Absolute OPD Demodulation

Absolute OPD demodulation is a universal method for readout of an FFP sensor, by which higher measurement accuracy could be obtained, compared with the intensity demodulation. As mentioned above, the measurand loaded on the FFP sensor is modulated spectrally; therefore, the interrogators are usually meant to measure the spectra and then convert the results to measured data [8]. Thus, getting the spectrum of the signal light is crucial. There are several common methods to obtain full spectra of FFP sensors including broadband light source combining with spectrometer, broadband light source combining with tunable filter, and swept or tunable laser combining with photodetector.

5.3.1 Spectral Interrogation Based on Spectrometers

The general schematic for full spectra acquiring of an FFP sensor using a spectrometer is demonstrated in Figure 5.4. The FFP sensor is illuminated by a broadband light source. The reflected light from

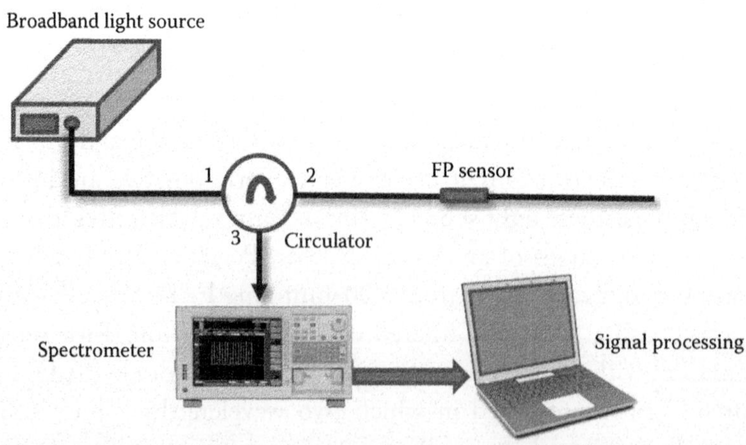

Figure 5.4 General schematic diagram of FFP spectra interrogation based on spectrometer.

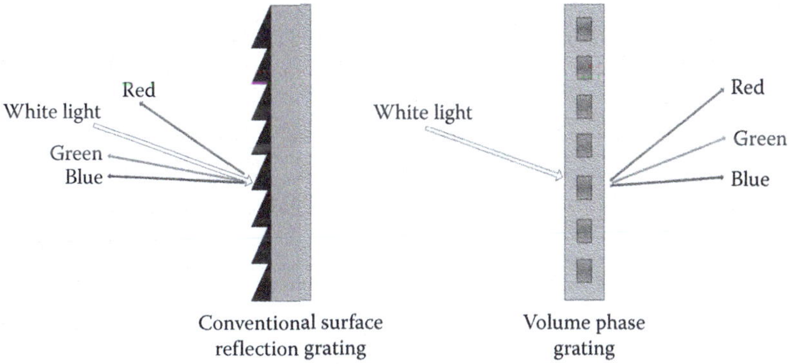

Figure 5.5 Operation schematic of diffraction gratings.

the sensor is received by a spectrometer. The light source could be SLED, SLD, or Erbium-doped fiber ASE light source, whose coherent lengths are generally shorter than the OPDs of FFP sensors.

A typical spectrometer is diffraction grating combining with photodetectors array [9,10]. Volume phase grating (VSP) is a kind of diffraction grating, like a conventional optical device used to spatially separate the different wavelengths or colors contained in a beam of light [10]. The device consists of a collection of diffracting elements (narrow parallel slits or grooves) separated by a distance comparable to the wavelength of light under test, as given in Figure 5.5. The VSP spectrometer is a highly integrated device and easy to mass produce, and has been successfully commercialized as a spectral interrogator for FFP sensors. The operating diagram of using VPG to measure the full spectrum of an FFP sensor is schematically shown in Figure 5.6 [10].

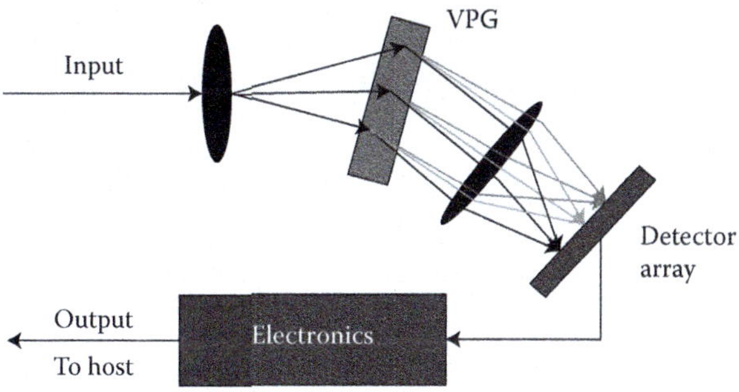

Figure 5.6 Schematic diagram of VPG spectrometer.

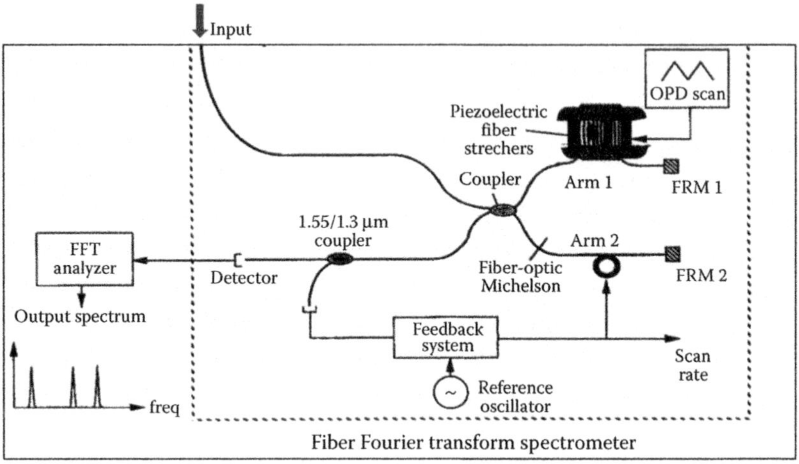

Figure 5.7 Schematic diagram of fiber Fourier transform spectrometer.

Another method to interrogate the spectra of FFP sensors is to use Fourier transform spectroscopy (FTS), as shown in Figure 5.7. The system uses a fiber-optic Michelson interferometer with one arm wrapped on a piezoelectric-driven stretcher capable of inducing a similar to 10-cm fiber length change to sample the spatial domain data. The spectrum can be obtained by Fourier transformation of the sampled data. Passive polarization compensation is utilized to eliminate the possibility of random polarization fading in the interferometer causing apodization of the interferogram. FTS is an efficient analytical tool for the analysis of broadband spectra with high sensitivity and reasonable resolution [11,12]. However, complex structure, low measuring speed, large volume, and high price limit its applications.

5.3.2 Spectral Interrogation Based on Tunable Filters

One of the most successful techniques for spectral interrogation of FFP sensors is based on the use of a tunable narrow bandpass filter [13], as shown in Figures 5.8 and 5.9. The commonly used filter is a fiber-pigtailed and piezo-electrical transducer (PZT)-driven tunable FP filter (TFPF) [14]. In Figure 5.8, the FFP sensor is illuminated by a broadband light source. The light reflected by the FFP sensor is

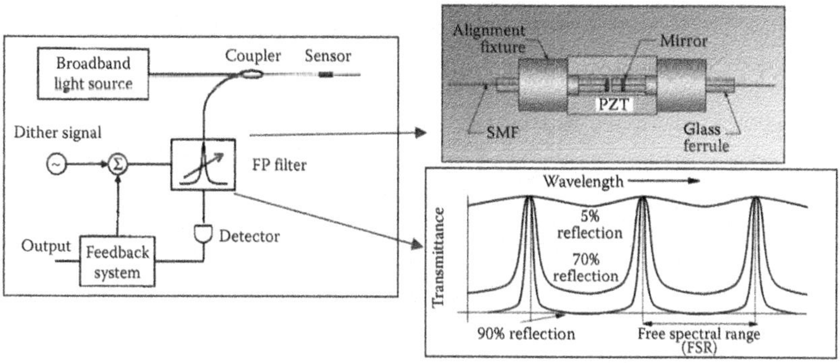

Figure 5.8 Schematic diagram of spectrum interrogation by arranging TFPF in front of detector.

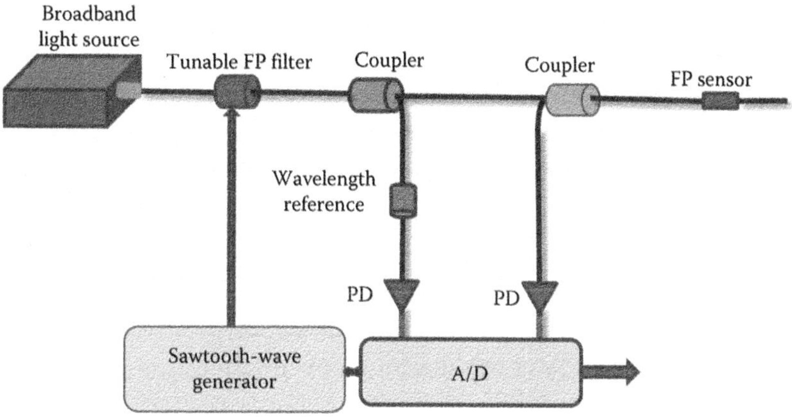

Figure 5.9 Schematic diagram of spectrum interrogation by arranging TFPF behind light source.

filtered by the TFPF to allow a narrowband light at a certain wavelength to be received by the detector. The transmission wavelength of the TFPF is controlled by an electrical driving signal. When the pass wavelength of the TFPF scans a wavelength range, the spectrum of the FFP sensor is interrogated. The scheme in Figure 5.9 could perform the same function in Figure 5.8. The only difference is that the TFPF is directly connected to the light source. If the power density of the light source is high, multiple parallel spectra interrogations could be realized by splitting the light after the TFFP. Spectral interrogation scheme based on the TFPF is low cost and simple, but it is difficult to achieve high signal-to-noise ratio and dynamic range due to the relatively low optical power after filtering.

5.3.3 Spectral Interrogation Based on Swept or Tunable Lasers

The scheme in Figure 5.9 could be theoretically used to interrogate multiple sensors, if the light intensity is high enough. In fact, power density of a broadband light source is relatively low, which is difficult to support multiple branches. A swept or tunable laser could be used to realize multiple sensor interrogation by power splitting due to its high optical intensity at a certain single wavelength, as shown in Figure 5.10. The coherent length of the laser is much longer than the OPD of the FFP sensor. The key point of such a scheme is how to construct a swept or tunable laser.

Generally, the swept laser is a ring fiber laser with a tunable FFP filter to select the output wavelength [14]. Erbium-doped fiber amplifier or semiconductor amplifier can be used as gain medium. When the light source scans a cycle in a certain wavelength range, each wavelength of light reflected from the FFP sensor is detected by a photodetector, which could be used to recover the spectrum of the sensor.

On the other hand, a semiconductor tunable laser is also used to obtain the full spectrum of an FFP sensor by scanning its output wavelength. Unlike the swept fiber ring laser, the output wavelength of a semiconductor tunable laser is shifted by changing its driving current to modulate the resonate wavelength of the laser cavity. The semiconductor laser could achieve much faster tuning speed and more stable laser output due to its shorter laser cavity [15].

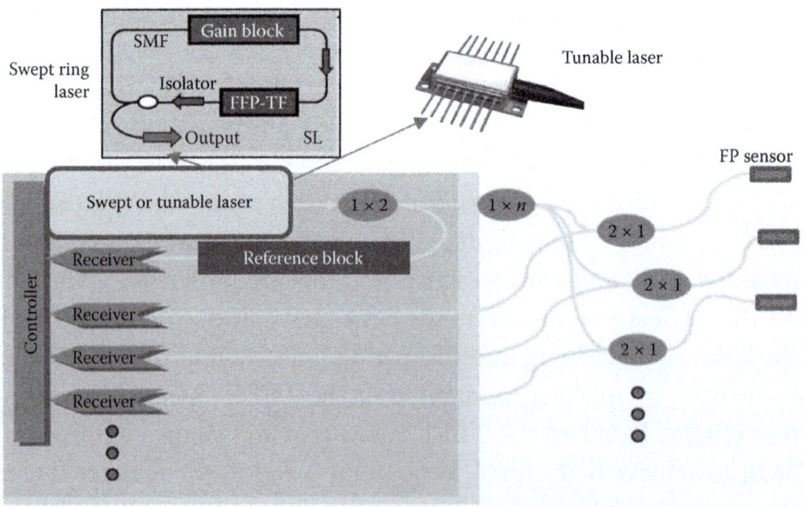

Figure 5.10 Spectra interrogation system based on swept or tunable laser.

Table 5.1 Comparison of Different Spectral Interrogation Methods

METHOD	TYPICAL RESOLUTION	WAVELENGTH RANGE (NM)	REFERENCES
Diffraction grating + CCD	0.14 –7 nm	400–1100 or 900–2500	9
VSP + CCD	1 pm	1510–1590	10
FTS	0.02 nm	600–1700	12
TFPF	0.02–0.2 nm	1460–1620	14
Swept laser	1 pm	1510–1590	14
Tunable laser	1 pm to 0.1 nm	1250–1360	15

The advantages of the scheme based on swept or tunable laser are easy to realize multiple parallel spectral interrogations with high dynamic range and SNR.

5.3.4 Comparison of Different Spectral Interrogation Methods

Table 5.1 gives a comparison of different spectral interrogation methods. It can be seen that spectral interrogators based on VSP, swept laser, or tunable laser could offer superior performance.

5.4 OPD Demodulation Methods

5.4.1 OPD Demodulation Based on Spectrum Interrogation and Fringe Counting

After obtaining the spectrum of FFP sensors using above-mentioned spectrum interrogation methods, the next mission is how to demodulate the OPDs of FFP sensors from their spectra. A simple method is called fringe counting [16–19]. For a certain fringe m as shown in Figure 5.11, since the peak wavelength corresponds to the minimum value of the cosine curve, we can get

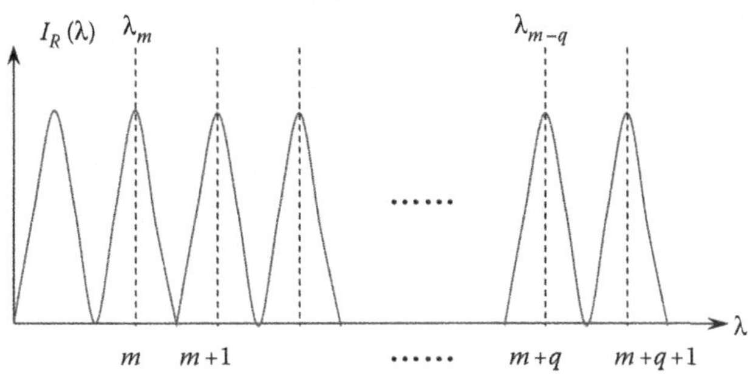

Figure 5.11 Schematic diagram of fringe counting.

$$\frac{4\pi nL}{\lambda_m} = 2m\pi + \pi \tag{5.1}$$

For the fringe $m - q$, we can get

$$\frac{4\pi nL}{\lambda_{m-q}} = 2(m-q)\pi + \pi \tag{5.2}$$

Therefore, the OPD of the FFP cavity is derived as

$$L = \frac{q}{2n}\left(\frac{\lambda_{m-q} + \lambda_m}{\lambda_{m-q} - \lambda_m}\right) q = 1, 2, 3, \ldots, \lambda_{m-q} > \lambda_m \tag{5.3}$$

Demodulation resolution using this method strongly depends on the peak or pit bandwidth and fringe number. Generally, in order to achieve high resolution, an FFP sensor with narrow peak or pit bandwidth and more fringes are highly desired.

To further improve the performance of demodulation, a combination of one-peak and two-peak tracking methods was proposed, which presented a nanometer-scale demodulation stability [18]. A two-step method including fringe counting and phase estimation was also proposed to improve demodulation performance [19]. A spectrum fitting algorithm was also demonstrated by Han et al. to achieve both high-resolution, absolute measurement of the cavity length, and a large dynamic measurement range simultaneously, but errors might be introduced by various imperfections of the sensor [20].

5.4.2 OPD Demodulation Based on Spectrum Interrogation and Fourier Transform

A universal method for the OPD demodulation is to use the fast Fourier transform (FFT) [21–28], since the reflected spectrum of the FFP cavity is a periodic curve in frequency domain. The reflected spectrum could be expressed as

$$I_R = D(k) + C \cdot \cos(Lgk + \varphi_0) \tag{5.4}$$

where $k = 2\pi/\lambda$ (λ is wavelength), L is the OPD, $D(k)$ is the light source background, C is the contrast factor, and φ_0 is the initial phase.

After Fourier transformation, Equation 5.4 changes to

$$G(n) = A(n) + B(n - n_0) + B^*(n + n_0) \tag{5.5}$$

where * denotes a complex conjugate and the upper-case letters denote the Fourier spectrum. Here, n is the digital frequency in the FFT spectrum. If the cavity length is long enough, the digital frequency n_0 is much larger than the spread of the spectrum caused by the variations of $D(\nu)$. For a general FFT method to interrogate the cavity length, the peak position of carrier frequency n_0 is detected, as shown in Figure 5.12, and then the optical path L can be calculated by using [27]

$$L = cn_0 / N\delta k \tag{5.6}$$

where δk is the optical spectrum sampling frequency interval, N is the points of FFT, and c is the light velocity in vacuum.

Because n_0 is an integer, one solution for achieving high OPD resolution is to interpolate massive data points using zero padding during FFT, which makes the density of the data points high enough so that the phase information can be retrieved from any given peak location. The drawback of this technique is the greatly increased amount of calculation since the zero padding usually needs to expand the length of the FFT data to 10 times or more to achieve high data point density for reducing the peak position reading error.

For improving demodulation performance, the real peak n_p could be calculated from the FFT data using the Buneman frequency

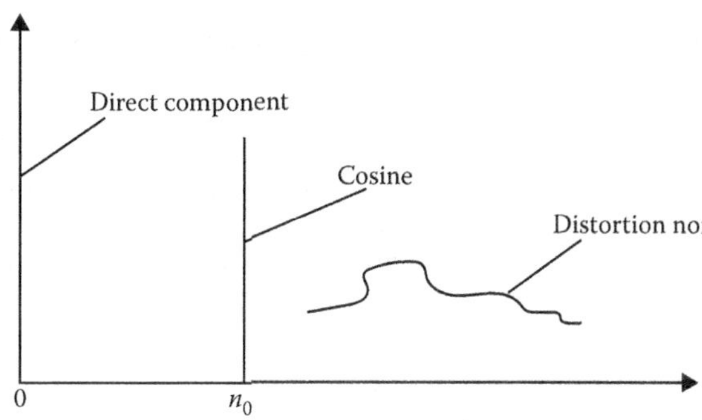

Figure 5.12 FFT spectrum of the FFP sensor.

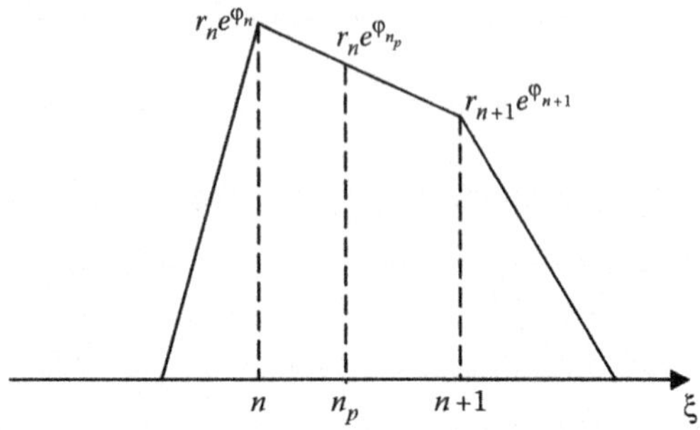

Figure 5.13 Schematic of the positive FFT peak of an envelope-removed interferometric fringe.

estimation, where n_0 is not an integer any more. As shown in Figure 5.13, the real peak n_p locates between the data points with index numbers of n and $n + 1$, and complex data values of $r_n e^{\varphi_n}$ and $r_{n+1} e^{\varphi_{n+1}}$.

n_p could be calculated by

$$n_p = n + \frac{N}{\pi} \arctan\left[\frac{\sin \pi/N}{\cos \pi/N + r_n/r_{n+1}}\right] \tag{5.7}$$

The optical path L can be calculated by using Equation 5.6, where n_p substitutes for n_0. Furthermore, from the first estimated n_p, type II peak index can be calculated by [26]

$$n_p^1 = \frac{1}{2k_0/k_1 - k_0 + 1}\left(\frac{\varphi_n - \varphi_0}{\pi} + n + 2\mathrm{INT}\left[\frac{k_0 n_p}{k_1 - k_0} + \frac{\varphi_0}{2\pi} + \frac{n_p - n}{2} - \frac{\varphi_n}{2\pi}\right]\right) \tag{5.8}$$

where k_0 and k_1 are the wave numbers of the spectrum's first and last data points, and INT is a rounding calculation. The optical path L can be calculated by using Equation 5.8, where n_p^1 substitutes for n_0. Based on Equation 5.8, the OPD demodulation performance is much better than direct peak tracking; 1-nm demodulation resolution and 10-kHz demodulation speed have been achieved [26].

A similar method based on the combination of the Fourier transform method and the minimum mean-square error estimation–based signal processing method was presented by Zhou and Yu [27], which is capable of providing subnanometer resolution and absolute measurement over a wide dynamic range.

Another approach to demodulate the OPD using the FFT is to select $B(n-n_0)$ by filtering, and to obtain the inverse Fourier transform of $B(n-n_0)$ as

$$\hbar(\lambda) = -\frac{1}{2}b(\lambda)\exp(j\varphi(\lambda)) \tag{5.9}$$

where $\varphi(\lambda) = 2\pi/\lambda L$. The imaginary part gives the principal value of the phase change with modulo 2π, and the phase $\varphi(\lambda)$ is wrapped into the range $[-\pi, +\pi]$. So, there are discontinuities with 2π phase jumps in $\varphi(\lambda)$. This wrapped phase is corrected by using a phase unwrapping algorithm. Then, a measurement of a large phase change exceeding 2π can be realized. The optical path d, can be calculated when the phase $\varphi(\lambda)$ is obtained. However, when we scan the wavelength from λ_0 to λ_1, we obtain a phase change $\Delta\varphi(\lambda)$, so the optical path L is actually obtained from the following equation [28]:

$$L = \frac{\lambda_1\lambda_0}{2\pi(\lambda_1 - \lambda_0)}\Delta\varphi(\lambda) \tag{5.10}$$

Based on Equation 5.10, in order to get better performance, the light source envelope should be removed to eliminate the low-frequency cross talk. Resolution of 10 nm could be realized using this method [28].

5.4.3 OPD Demodulation Based on Interferometric Scanning in WLI

Besides the above-mentioned OPD demodulation method, an interferometric scanning "WLI" combining with a broadband light source could also be applied to determine the optical path lengths of FFP sensors. The implication of the term "white light" is that the spectrum is sufficiently broad that the coherence length of the light source is much less than the round-trip optical path length of the FFP. Suitable light sources for WLI include SLDs, ASE source, and tungsten lamps. These sources typically have spectral widths of several tens of nanometers.

A WLI system for reading out an FFP sensor is illustrated schematically in Figure 5.14a, while the optical monitoring

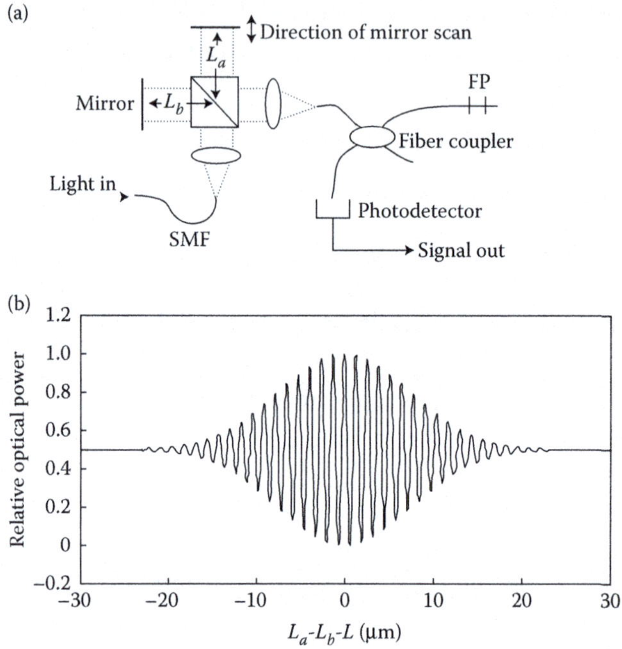

Figure 5.14 Measurement of interferometer phase shift using low-coherence (white light) interferometry: (a) optical configuration and (b) fringe pattern.

signal is shown in Figure 5.14b. Light from the broadband source is transmitted or reflected by both the sensing and reference interferometers before reaching the photodetector [29]. When the optical path length of the reference interferometer is scanned, the photodetector output is a fringe (interference) pattern that has its maximum amplitude (peak of central fringe) where the OPD of the reference interferometer and that of the sensor interferometer are equal. The width of the pattern is proportional to the coherence length of the light source. The most common reference interferometer is the Michelson, although the Mach–Zehnder [30] and the FP [31] have also been used.

The fringe data as shown in Figure 5.14 are processed electronically to determine the exact OPD of the scanned reference interferometer corresponding to the central fringe peak. Identification of the central fringe is a key issue. A mistaken central fringe identification leads to an error in round-trip optical path length of at least one wavelength. An approach that greatly reduces the chance of central fringe error by

Table 5.2 Methods for Demodulation of Optical Path Difference (OPD)

METHOD	RESOLUTION	REFERENCES
Fringe counting	Depends on fringe 3 dB bandwidth and numbers	[16,18]
Combined one- and two-peak fringe counting	1.5 nm	[17]
Fringe counting and spectrum matching	<1 nm	[19]
Spectrum fitting	0.288 nm; errors will be increased by imperfections of sensors and model	[20]
FFT peak search	1–50 nm; depends on spectrum wavelength range and resolution	[21,22,25]
FFT peak estimation	1 nm	[24,26,27]
FFT WLI	±10 nm	[28]
Interferometric scanning WLI	Depends on scanning wavelength range	[29–31]

improving the fringe amplitude contrast uses two broadband sources with a wide wavelength separation [32]. The drawback of this method is that the OPD demodulation speed is relatively slow, limited by the scanning speed of the mirror.

5.4.4 Comparison of OPD Demodulation Methods

Comparison of OPD demodulation methods is given in Table 5.2. FFT-based demodulation method is helpful to obtain good demodulation performance.

5.5 Multiplexing Methods of FFP Sensors

5.5.1 Spatial Frequency-Division Multiplexing [33]

A multiplexing technique that allows for FFP sensors with different lengths to be arranged parallely by using an optical coupler is given in Figure 5.15. These sensors share a broadband light source and a

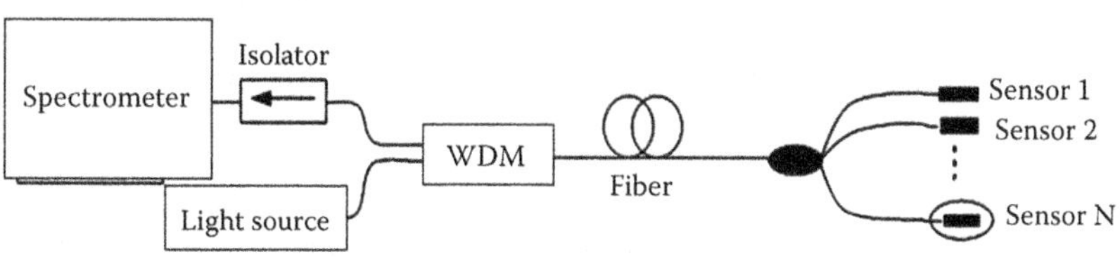

Figure 5.15 Schematic configuration of SFDM of FFP sensors.

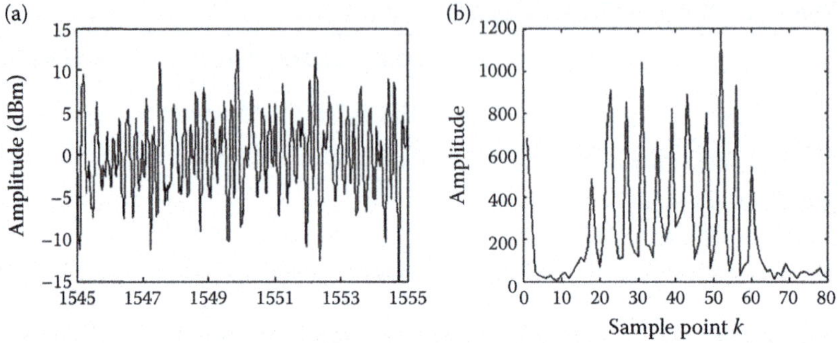

Figure 5.16 (a) Mixed signal of 11 FFP sensors and (b) FFT spectrum of 11 signals.

spectrometer. Their mixed reflective optical signals are received by the spectrometer, then processed using FFT. These sensors with different cavity lengths are located at different frequency bands in the FFT spectrum, which could be clearly differentiated. Therefore, they could be individually demodulated using FFT-based OPD demodulation methods.

A series of 11 FFP sensors is multiplexed in 1550-nm band. Their mixed optical signal and FFT spectrum are given in Figure 5.16. It can be seen that the spatial-frequency component corresponding to each cavity length can be distinguished clearly so that multiplexing of up to 11 FFP sensors is achievable.

Such a spatial frequency-division multiplexing (SFDM) sensor system with the advanced demodulation algorithm can be used for a wide range of applications, in particular, for those applications where remote monitoring with a large number of sensors is essential. By combining with optical switches, the total sensor number, which could be multiplexed, may be up to several hundreds. Also, it can reduce the cost of the whole sensor system to a very large extent and have excellent performance-to-price ratio due to the significant improvement of the multiplexing capability over conventional FFP sensor systems.

5.5.2 Coarse Wavelength-Division Multiplexing [34]

A multiplexing technique based on coarse wavelength-division multiplexing (CWDM) could be used to separate different

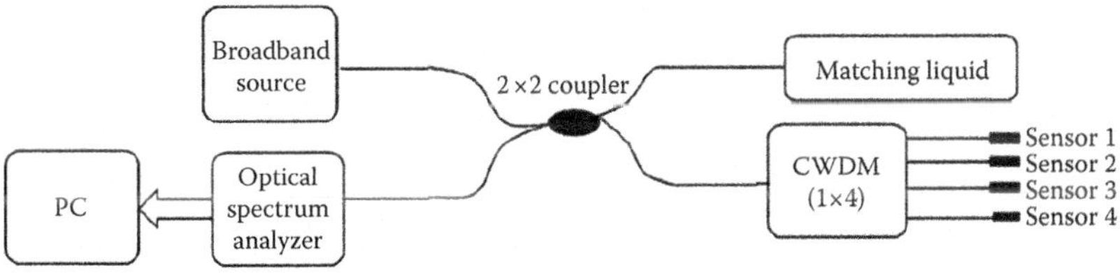

Figure 5.17 Example of multiplexing four sensors using CWDM.

sensors into their individual wavelength bands. These sensors share a broadband light source and a spectrometer, but do not interfere with each other. As shown in Figure 5.17, an example of multiplexing four sensors is given by a 1 × 4 CWDM coupler. It can be seen clearly from Figure 5.18 that the four signals corresponding to the four FFP sensors are in different wavelength regions and can be demodulated with similar ways as a single FFP sensor.

5.5.3 SFDM/CWDM of FFP Sensors [35]

Furthermore, CWDM and SFDM are combined to enhance the multiplexing capability considerably. The schematic diagram is given by Figure 5.19. Multiplexing of more than 10 FFP sensors was experimentally demonstrated using SFDM in each wavelength channel predetermined by CWDM. This method can multiplex a large number of sensors of up to 100 in principle and hence can reduce the cost of

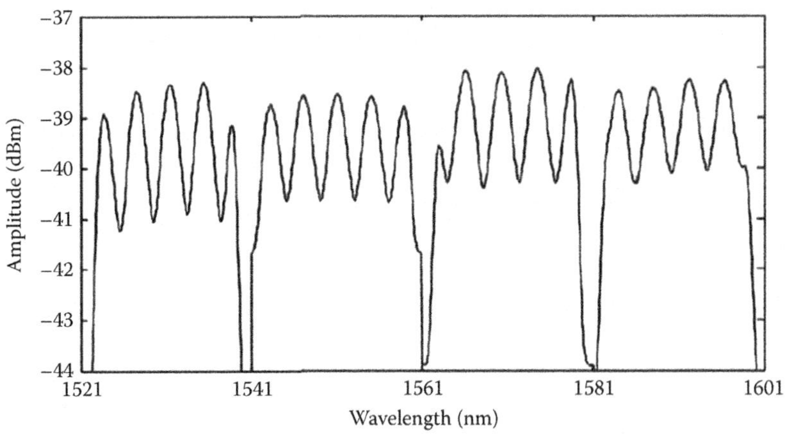

Figure 5.18 Reflective spectrum of the FFP sensors separated by CWDM.

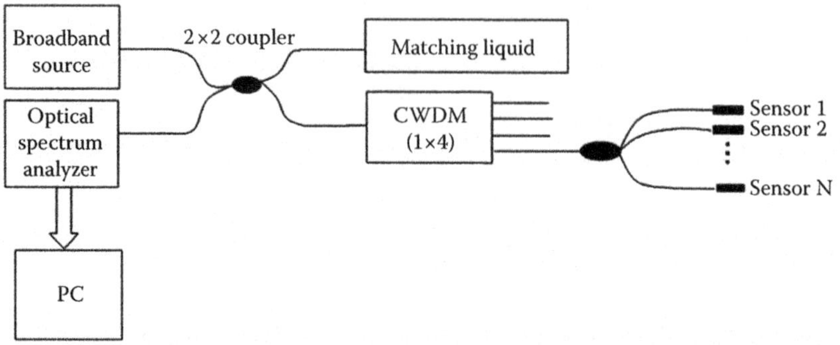

Figure 5.19 Schematic diagram of hybrid SFDM/CWDM multiplexing.

the whole sensor system greatly. Also, such a hybrid method can take advantages of both SFDM and CWDM when applied to the multiplexing of a large number of sensors, that is, excellent multiplexing capability when using SFDM, and good efficiency in using every part of the optical power of the broadband source with CWDM.

As a verification example, the mixed spectrum of four FFP strain sensors is shown in Figure 5.20a. By using filtering and FFT, the waveform of the mixed signal is shown in Figure 5.20b. It can be seen that the spatial-frequency component corresponding to each cavity length of these four FFP sensors can be distinguished clearly.

5.5.4 Time-Division Multiplexing

The reflective FFP sensor is suited for time-division multiplexing using a pulsed light source such as semiconductor LD [36].

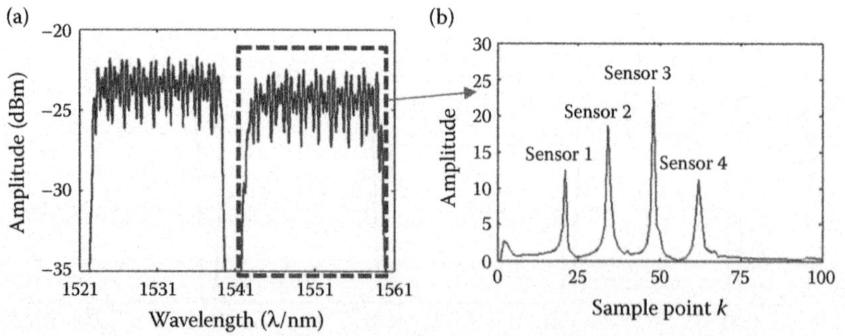

Figure 5.20 (a) Reflective signal from two CWDM channels and (b) FFT spectrum of four FFP sensors with different cavity lengths.

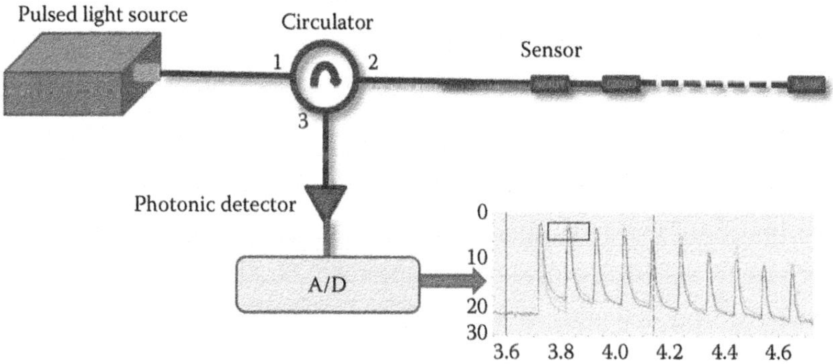

Figure 5.21 Schematic diagram of TDM system for FFP sensors.

A schematic diagram is given in Figure 5.21. In such a system, fiber delay lines with different lengths are provided between the transmitter and receiver such that, for each laser pulse, the photodetector receives reflected pulses from each of the sensors in different time slots. The signals can be processed by digital means under microprocessor control. In such a scheme, the reflected waveforms are sampled for analog-to-digital conversion at fixed time delays relative to the start of the pulse, and the samples are averaged digitally. In-phase and quadrature signals from the interferometers needed for high-sensitivity sensing can be obtained by adjusting the dc bias current of the laser between pulses. A reference FFP time multiplexed with the sensors can be used to correct the fluctuations of the light source and drift of the laser wavelength.

To minimize cross talk between sensors, they are arranged in a single fiber. The reflectivity of the sensor should be controlled at a low level (typical <4%). However, cross-talk effects are limited as the transmissions vary at the individual sensors. This cross talk can be further eliminated by measuring the Rayleigh back-scattered level just in front of the sensor, which can be used as the reference for calculating the reflectivity of a particular sensor in the network [37]. The alternative method for cross-talk elimination can be summing of the losses on all preceding sensors, while calculating the reflectivity of the observed one.

A fatal drawback of this technique is that it puts forward a tight requirement on the consistency of different sensors, for achieving the same quadrature points.

5.5.5 Multiplexing of FFP Sensors Based on Microwave-Assisted Reconstruction

A method of multiplex fiber-optic interferometers for distributed sensing through microwave-assisted separation and reconstruction of optical interferograms in the spectrum domain is proposed [38]. Cascaded FFP sensors could be demodulated individually using this technique. The approach is schematically shown in Figure 5.22. The light from a broadband source is launched into a tunable optical filter and then intensity-modulated by a microwave signal whose modulation frequency can be scanned via computer control. The microwave-modulated light, where the optical signal is the carrier and the microwave is the envelope, is then sent into an optical fiber with cascaded reflective FFP sensors through a fiber-optic circulator. The reflection of each FFP sensor can be designed to be weak enough so that the light can be transmitted over many sensors and the multiple reflections within each sensor is negligible. The coherence length of the light, which is determined by the bandwidth of the tunable filter, is much larger than the OPD of the FFP sensors but much smaller than the distance between two adjacent

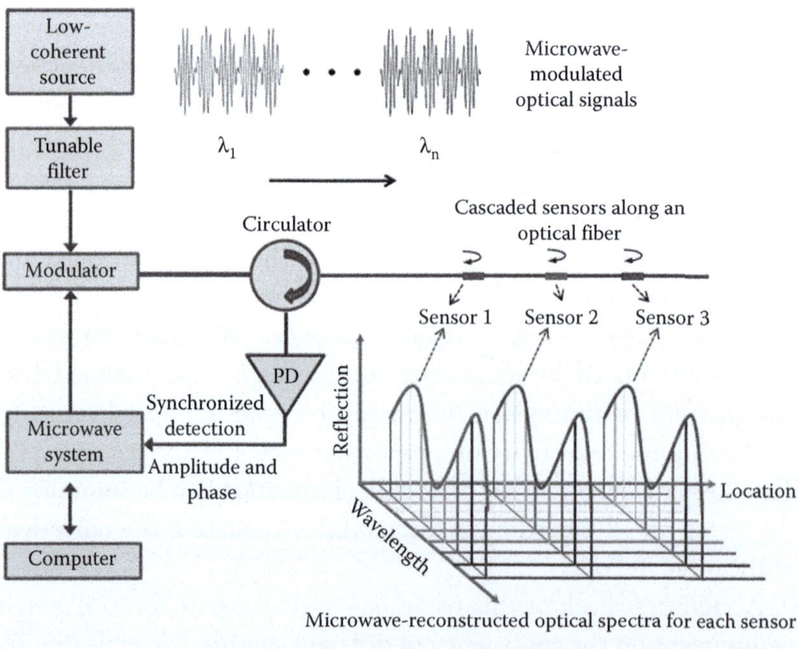

Figure 5.22 Schematic illustration of the microwave-assisted multiplexing of FFP sensors.

sensors. As a result, optical interference occurs within one sensor while the optical interference between sensors is avoided.

The optical detection is synchronized with the microwave modulation frequency so that the amplitude and the phase are obtained. After scanning the microwave frequency through the entire available range, the complex microwave reflection spectrum (with both amplitude and phase) is obtained. By applying a complex and inverse FFT to the microwave spectrum, a series of delta functions are obtained at discrete time positions. The discrete time domain signals are proportional to the optical interference signals of the cascaded FFP sensors at a particular optical wavelength (λ_m) determined by the tunable filter, and also indicate the position of the sensors along the fiber.

By sweeping the optical wavelength and repeating the microwave measurement, the discrete optical interference signals (separated in the time domain) at different wavelengths are obtained. These data points can then be used to construct the optical interferograms of the cascaded FFP sensors. The concept of multiplexing of FFP sensors based on microwave-assisted reconstruction is like multi-wavelength time-division multiplexing (TDM) to some extent. Compared with TDM using laser pulses, it could achieve higher spatial resolution and obtain the full spectrum of sensors, and does not require all the sensors with the same OPD. The disadvantage of this approach is time consumption.

5.5.6 Coherence Multiplexing Based on Interferometric Scanning WLI

The WLI monitoring scheme can be extended to the interrogation of FFP sensors with different lengths in a sensor network, as illustrated in Figure 5.23. This scheme is known as coherence multiplexing (CM) [39]. CM also requires the use of a reference interferometer matched in length to within the coherence length of the light source (about 10 μm for a typical LED or SLD) to the sensor FFP being interrogated. Light from the broadband source must be transmitted or reflected by both the sensor and reference interferometers before reaching the photodetector. If the reference interferometer is scanned, one fringe peak is observed. In one experiment using a quartz halogen lamp as the light source and a scanned Michelson reference interferometer, six FFP strain sensors with different optical cavity lengths are multiplexed in series along the length of an SMF [40]. In another

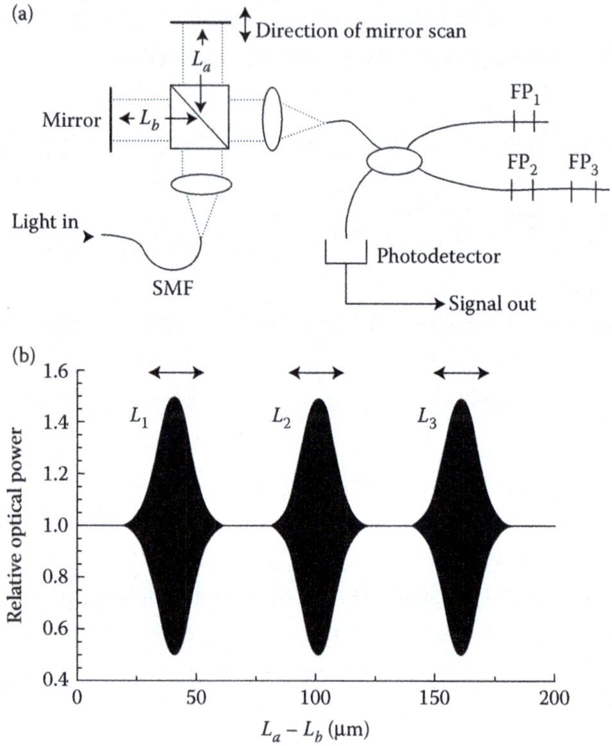

Figure 5.23 Arrangement for CM: (a) optical system for sensor interrogation and (b) calculated fringe patterns from three FFP interferometers of lengths L_1, L_2, and L_3.

case, two FFP strain sensors are multiplexed in serial and parallel arrangements using a multimode LD as the light source [41].

5.5.7 Comparison of Multiplexing Methods

Comparison of multiplexing methods is given in Table 5.3. It can be seen that different methods have different advantages and disadvantages.

Table 5.3 Comparison of Different Multiplexing Methods

METHODS	CAPACITY	REFERENCES
SFDM	Medium	33
CWDM	Low	34
SFDM/CWDM	High	35
TDM	Medium, depends on sensor insertion loss	36,37
Microwave-assisted reconstruction	Medium, depends on sensor insertion loss	38
Interferometric scanning WLI	Low	39–41

References

1. Udd, E. 2002. *Overview of Fiber Optic Sensors* (Vol. 1). New York: Marcel Dekker.
2. Lu, E., Ran, Z., Peng, F., Liu, Z., and Xu, F. 2012. Demodulation of micro fiber-optic Fabry–Perot interferometer using subcarrier and dual-wavelength method. *Optics Communications*, 285(6), 1087–1090.
3. Murphy, K. A., Gunther, M. F., Vengsarkar, A. M., and Claus, R. O. 1991. Quadrature phase-shifted, extrinsic Fabry–Perot optical fiber sensors. *Optics Letters*, 16(4), 273–275.
4. Potter, J., Ezbiri, A., and Tatam, R. P. 1997. A broad band signal processing technique for miniature low-finesse Fabry–Pérot interferometric sensors. *Optics Communications*, 140(1), 11–14.
5. Furstenau, N. and Schmidt, M. 1998. Interferometer vibration sensor with two-wavelength passive quadrature readout. *IEEE Transactions on Instrumentation and Measurement*, 47(1), 143–147.
6. Lo, Y. L., Sirkis, J. S., and Chang, C. C. 1997. Passive signal processing of in-line fiber etalon sensors for high strain-rate loading. *Journal of Lightwave Technology*, 15(8), 1578–1586.
7. Schmidt, M. and Fürstenau, N. 1999. Fiber-optic extrinsic Fabry–Perot interferometer sensors with three-wavelength digital phase demodulation. *Optics Letters*, 24(9), 599–601.
8. Taylor, H. F. 2002. Fiber optic sensors based upon the Fabry–Perot interferometer. *Fiber Optic Sensors*, 20020456.
9. http://www.oceanoptics.cn
10. http://www.bayspec.com/
11. Davis, M. A. and Kersey, A. D. 1995. Application of a fiber Fourier transform spectrometer to the detection of wavelength-encoded signals from Bragg grating sensors. *Journal of Lightwave Technology*, 13(7), 1289–1295.
12. http://tmi.yokogawa.com
13. Kersey, A. D., Berkoff, T. A., and Morey, W. W. 1993. Multiplexed fiber Bragg grating strain-sensor system with a fiber Fabry–Perot wavelength filter. *Optics Letters*, 18(16), 1370–1372.
14. http://www.micronoptics.com
15. http://www.auniontech.com
16. Bhatia, V., Sen, M. B., Murphy, K. A., and Claus, R. O. 1996. Wavelength-tracked white light interferometry for highly sensitive strain and temperature measurements. *Electronics Letters*, 32(3), 247–249.
17. Qi, B., Pickrell, G. R., Xu, J., Zhang, P., Duan, Y., Peng, W. et al. 2003. Novel data processing techniques for dispersive white light interferometer. *Optical Engineering*, 42(11), 3165–3171.
18. Jiang, Y. 2008. High-resolution interrogation technique for fiber optic extrinsic Fabry–Perot interferometric sensors by the peak-to-peak method. *Applied Optics*, 47(7), 925–932.
19. Huang, H. 2006. Data interrogation for Fabry–Perot white light interferometry. In *Smart Structures and Materials* (pp. 617419–617419), San Diego, CA. International Society for Optics and Photonics.

20. Han, M., Zhang, Y., Shen, F., Pickrell, G. R., and Wang, A. 2004. Signal-processing algorithm for white-light optical fiber extrinsic Fabry–Perot interferometric sensors. *Optics Letters*, 29(15), 1736–1738.
21. Takeda, M., Ina, H., and Kobayashi, S. 1982. Fourier-transform method of fringe-pattern analysis for computer-based topography and interferometry. *JOSA*, 72(1), 156–160.
22. Liu, T. and Fernando, G. F. 2000. A frequency division multiplexed low-finesse fiber optic Fabry–Perot sensor system for strain and displacement measurements. *Review of Scientific Instruments*, 71(3), 1275–1278.
23. Boulet, C., Hathaway, M., and Jackson, D. A. 2004. Fiber-optic-based absolute displacement sensors at 1500 nm by means of a variant of channeled spectrum signal recovery. *Optics Letters*, 29(14), 1602–1604.
24. Shen, F. and Wang, A. 2005. Frequency-estimation-based signal-processing algorithm for white-light optical fiber Fabry–Perot interferometers. *Applied Optics*, 44(25), 5206–5214.
25. Majumdar, A. and Huang, H. 2008. Development of an in-fiber white-light interferometric distance sensor for absolute measurement of arbitrary small distances. *Applied Optics*, 47(15), 2821–2828.
26. Yu, Z. and Wang, A. 2015. Fast white light interferometry demodulation algorithm for low-finesse Fabry–Perot sensors. *IEEE Photonics Technology Letters*, 27(8), 817–820.
27. Zhou, X. and Yu, Q. 2011. Wide-range displacement sensor based on fiber-optic Fabry–Perot interferometer for subnanometer measurement. *IEEE Sensors Journal*, 11(7), 1602–1606.
28. Jiang, Y. 2008. Fourier transform white-light interferometry for the measurement of fiber-optic extrinsic Fabry–Perot interferometric sensors. *IEEE Photonics Technology Letters*, 2(20), 75–77.
29. Beheim, G. 1985. Remote displacement measurement using a passive interferometer with a fiber-optic link. *Applied Optics*, 24(15), 2335–2340.
30. Choi, H. S., Taylor, H. F., and Lee, C. E. 1997. High-performance fiber-optic temperature sensor using low-coherence interferometry. *Optics Letters*, 22(23), 1814–1816.
31. Lee, C. E. and Taylor, H. F. 1991. Fiber-optic Fabry–Perot temperature sensor using a low-coherence light source. *Journal of Lightwave Technology*, 9(1), 129–134.
32. Chen, S., Grattan, K. T. V., Meggitt, B. T., and Palmer, A. W. 1993. Instantaneous fringe-order identification using dual broadband sources with widely spaced wavelengths. *Electronics letters*, 29(4), 334–335.
33. Rao, Y. J., Jiang, J., and Zhou, C. X. 2005. Spatial-frequency multiplexed fiber-optic Fizeau strain sensor system with optical amplification. *Sensors and Actuators A: Physical*, 120(2), 354–359.
34. Zhou, C. X., Rao, Y. J., and Jiang, J. 2005. A coarse wavelength-division-multiplexed extrinsic fiber Fabry–Perot sensor system. In *Photonics Asia 2004* (pp. 219–224), Beijing, China. International Society for Optics and Photonics.

35. Rao, Y. J., Zhou, C. X., and Zhu, T. 2005. SFDM/CWDM of fiber-optic Fizeau strain sensors. *IEEE Photonics Technology Letters,* 17(6), 1259–1261.
36. Lee, C. E. and Taylor, H. F. 1988. Interferometric optical fibre sensors using internal mirrors. *Electronics Letters*, 24(4), 193–194.
37. Cibula, E. and Donlagic, D. 2007. In-line short cavity Fabry–Perot strain sensor for quasi distributed measurement utilizing standard OTDR. *Optics Express*, 15(14), 8719–8730.
38. Huang, J., Hua, L., Lan, X., Wei, T., and Xiao, H. 2013. Microwave assisted reconstruction of optical interferograms for distributed fiber optic sensing. *Optics Express*, 21(15), 18152–18159.
39. Davis, C. M., Zarobila, C. J., and Rand, J. D. 1988. Fiber-optic temperature sensor for microwave environments. In *1988 Los Angeles Symposium—OE/LASE'88* (pp. 114–119), Los Angeles, CA. International Society for Optics and Photonics.
40. Tuck, C. J. and Fernando, G. F. 1999. Multiplexed optical fiber Fabry–Perot sensors for strain metrology. In *1999 Symposium on Smart Structures and Materials*, Newport Beach, CA (pp. 322–329). International Society for Optics and Photonics.
41. Kaddu, S. C., Collins, S. F., and Booth, D. J. 1999. Multiplexed intrinsic optical fibre Fabry–Perot temperature and strain sensors addressed using white-light interferometry. *Measurement Science and Technology*, 10(5), 416.

6
Applications

When fiber optics was being developed for communications, efforts were underway to use the new technology for sensing at the same time. Optical fiber sensors are being broadly advanced as they have numerous advantages over conventional sensors such as the ability to function in harsh environments, small sizes, with electromagnetic interference, and many more. Among them, interferometer-based fiber-optic sensors have been implemented for a wide range of applications since 1980 [1]. After the introduction of FFPI, optical sensors based on FFPI have been extensively studied due to their high performance and ability for signal "amplification" (i.e., resonance). Also, FFPIs are attractive because of high sensitivity, large dynamic range, fast response for measurement of various parameters, such as temperature, strain, pressure, displacement, electrical/magnetic field, refractive index, flow rate, etc. and can also be used as embedded sensors in composite materials [2,3]. There are several ways that are being followed, each of which can be the link between the variable being sensed and an optical resonance of the FFPI: (1) linear expansion of a spacer length, (2) refractive index of the medium between reflectors, (3) reflector absorption or reflectivity, and (4) spectral absorption or scattering in the medium between reflectors. Hence, FFPIs have shown considerable potential for sensing in various fields, such as energy, aerospace, biomedicine, et al., which are discussed below.

6.1 Structural Health Monitoring

Structural health monitoring needs technologies of real-time monitoring and nondestructive detection, which are difficult to achieve using conventional methods. Numerous advantages of FFPI sensors, such as immunity to electromagnetic interference, less intrusive size, and greater resistance to corrosion, have made them attractive for structural health monitoring of infrastructures including bridges, buildings, dams, et al.

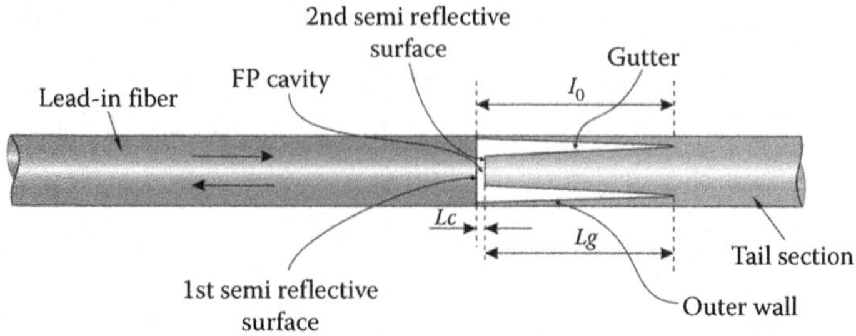

Figure 6.1 FP pressure sensor.

Engineering applications of FFPI sensors have been demonstrated by Gahan and Donlagic and their coworkers [4,5], respectively. Subsequently, in 2007, Needham invented an FFPI sensor with an air gap [6]. It can use a linear array processor for signal processing in industrial applications, such as gas turbines, engines, pressure vessels, pipelines, buildings, etc., to provide information on pressure, temperature, strain, vibration, or acceleration. An EFFPI strain sensor with sensitivity enhancement was developed for measuring strain of steel plate, as described in [7,8]. The proposed sensor is shown in Figure 6.1. The sensor is composed of a lead-in fiber that also forms the first EFFPI semi-reflective surface, an outer (semi-conical) cladding, a second EFFPI semi-reflective surface, a gutter that surrounds the second EFFPI semi-reflective surface, and a tail fiber that can be of arbitrary length. The strain sensitivity can be improved significantly as the gauge length L_g is much longer than the cavity length L_c.

Bridge's health monitoring also can be achieved based on EFFPI sensors [9]. Short-length EFFPI strain sensors, which are insensitive to the change of ambient temperature, have been successfully deployed on the concrete-based Hongcaofang Crossroads Bridge in Chongqing, China, as shown in Figure 6.2, which is 210 m long over seven spans. A set of four EFFPI strain sensors were attached to the centers of two spans to measure the static strain of the bridge. The results in Figure 6.3 indicate that this concrete bridge expands with the increase in temperature in daytime and contracts at night due to the temperature drop.

The strain peaks and troughs are just in accordance with the values of the temperature at 3:00 p.m. and 4:00 a.m., respectively. In order to evaluate the accuracy and repeatability of the EFFPI strain sensor,

Figure 6.2 Photograph of the Hongcaofang Crossroads Bridge.

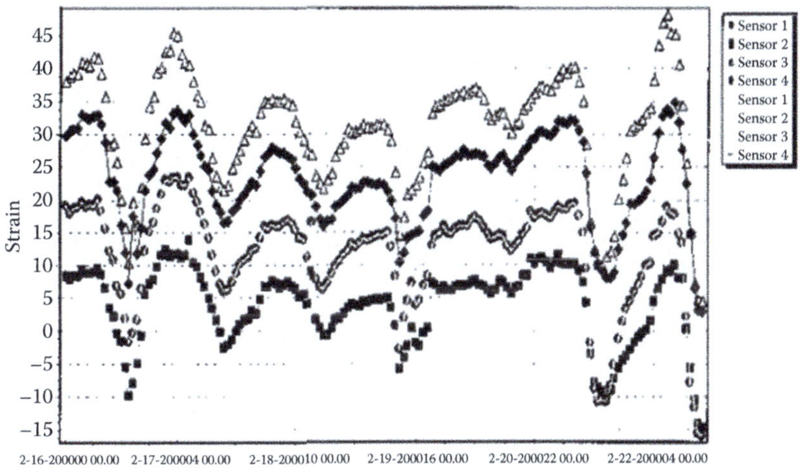

Figure 6.3 Results of the EFFPI strain sensor.

two experiments have been carried out based on a standard cantilever calibration setup with the standard electrical strain gauge as a reference. The results showed that the accuracy and repeatability of the EFFPI strain sensor were around ±1με.

6.2 Energy (Oil, Gas, Electricity)

Energy, especially of the non-regenerated types such as oil and gas, plays an essential role in our daily life. In 2005, an EFFPI sensor for use in oil downhole was reported [10]. The sensor comprises a hollow tube with one end bonded to the end face of optical fiber, and the

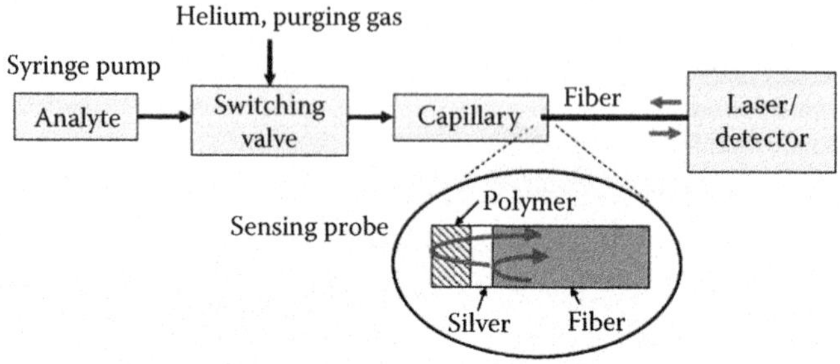

Figure 6.4 Structure of an EFFPI gas sensor.

other end bonded to a diaphragm. The fiber end face and diaphragm define an FP etalon whose cavity length varies when pressure or force is applied to a diaphragm. It can detect both pressure and acceleration, and has the advantages of being compact, inexpensive, and insensitive to temperature changes.

In 2009, Jing Liu proposed a gas sensor based on the EFFPI structure [11], which is shown in Figure 6.4. A versatile, easily fabricated EFFPI gas sensing probe has been developed, which can accommodate any polymers regardless of their reflective indexes. The sensing probe is illustrated in the enlarged part of Figure 6.4. It is composed of two layers: a silver layer and a vapor-sensitive polymer layer. The propagation of light in SMF will be partially reflected at the silver layer and the polymer–air interface. The interference spectrum is generated by these two reflected beams. When the sensing probe is exposed to analyte vapor, the vapor-sensitive polymer layer will interact with analyte, and the change of its RI or thickness will change the light path, which in turn causes the shift of the interference spectrum. By artificially introducing this reflective silver layer, polymer with various RI can be coated, thus tailoring the gas sensor for versatile usage.

Figure 6.5 shows another structure of an EFFPI pressure sensor for oil and gas downhole applications [12]. The air-gap (G) response due to the pressure is typically very linear and predictable. It is a function of the capillary tube material characteristics and the tube dimensions such as inner and outer diameters and gage length (L_0 in Figure 6.5). The air-gap change, ΔG, due to an applied pressure P can be expressed as

$$\Delta G = \frac{L_0 \cdot (P - P_0) \cdot r_0^2}{E \cdot \left(r_0^2 - r_i^2\right)} \cdot (1 - 2\mu) \tag{6.1}$$

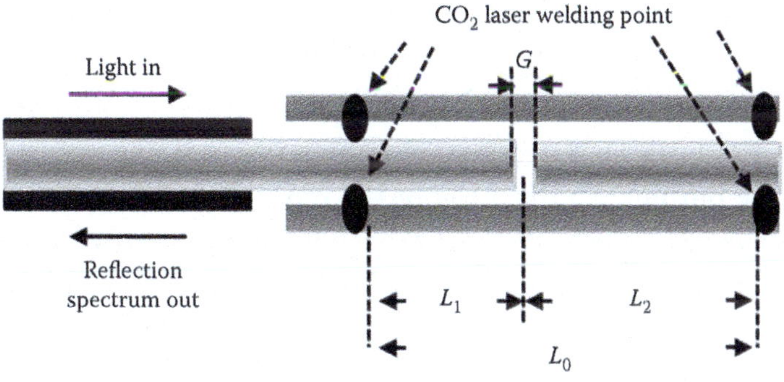

Figure 6.5 Structure of an EFFPI pressure sensor for oil and gas well applications.

where L_0 is the sensor gage length between the two bonding points, P_0 the pressure inside the tube, which is approximately the atmospheric pressure, E the Young's modulus of the glass material used for the capillary tube, r_o and r_i are outer and inner radius of the glass tube, and μ the Poisson's ratio of the glass. The air gap G is typically very small in comparison with the gage length L_0. As long as the thermal expansion coefficients (TEC) for the glass tube and the fibers are matched, the temperature cross-sensitivities (TCS) for this type of sensors can be negligible. The TEC for an optical fiber is quite close to the fused silica, which is around $0.5 \times 10^{-7}/°C$. Depending on the doping material and the doping level of its core and the size of the core, the TEC slightly varies for different fibers. TCS or $\Delta G/\Delta T$ can be expressed as

$$\frac{\Delta G}{\Delta T} = \alpha_0 \cdot L_0 - (\alpha_1 \cdot L_1 + \alpha_2 \cdot L_2) \tag{6.2}$$

where α_0, α_1, α_2 are the TEC of the capillary tube, the input fiber, and the reflection fiber, and L_1, L_2 are the lengths of the input fiber and the reflection fiber, respectively, as shown in Figure 6.5. Here,

$$L_0 = L_1 + L_2 + G, \quad G \ll L_0 \tag{6.3}$$

Because the air gap is very small in comparison with the gage length, and the TEC of the capillary tube is quite close to the TEC of the fibers, the sensor inherently has very small TCS at the atmospheric pressure. However, the TEC of the capillary tube is expected to increase under a higher pressure range. In other words, the TCS would become worse. The higher the pressure, the worse the TCS would be.

To avoid the TCS aggravation at high pressure, the sensor design could be optimized. The TEC of the input fiber cannot be adjusted as usually common SMF is used, but any kind of fiber with the same cladding diameter can be used as the reflection fiber. Also, the ratio between the lengths of the input and the reflection fibers can be adjusted to achieve a minimized total effect of TCS.

It is expected that the TEC of a fiber is higher if the doping level is higher and the core size is larger. When this kind of fiber is used for the reflection fiber, the TCS, $\Delta G/\Delta T$, would be a greater negative value at the atmospheric pressure as indicated in Equation 6.2. At a high pressure by increasing the TEC of the capillary tube, the TCS can be compensated to a smaller value, which could be zero at a certain point within the pressure range. If the TCS at the atmospheric pressure is reasonably controllable, the zero TCS point within the pressure range can be selectable, depending on the pressure range of interest.

6.3 Aerospace (Aircrafts, Engines)

There is a strong interest in the aerospace community to increase the safety and reliability of aircraft while reducing operating and maintenance costs by replacing conventional wire-sensing systems with fiber-optics sensors. Research on structural health monitoring of aerospace vehicles and aircrafts have been reported in 2000 [13]. Fiber-optic sensors have remarkable advantages in meeting the miniature size and weight requirements because of their improved multiplexing capability and their benefits including low cost and immunity to electromagnetic interference and lightning. One important type of sensor is a binary switch, which is actuated when the pilot presses a button [14]. The signal from the switch is "0" when no force is applied, and "1" when the operator-induced strain in the sensing element exceeds a given threshold. As shown in Figure 6.6, the white-light interferometry (WLI) monitoring setup consists of two bulk-scanned Michelson interferometers sharing a common mirror translation stage. One interferometer uses a super-luminescent diode

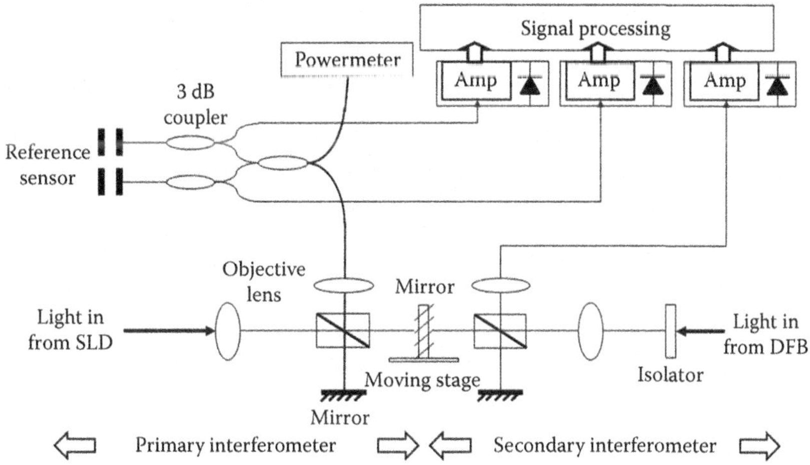

Figure 6.6 Setup of WLI monitoring system.

(SLD) operating at central wavelength of 1325 nm as the light source for monitoring a sensing FFPI and a reference FFPI, while the second interferometer uses a distributed feedback (DFB) laser light source at a 1540 nm peak emission wavelength. The output light from the DFB interferometer is detected, and the resulting fringe pattern is processed to determine the position of the stage with high precision. One complete fringe corresponds to a displacement of $\lambda/2$ at the DFB laser wavelength of 1540 nm.

The FFPI contains two dielectric internal mirrors forming an FP cavity of length L within a single-mode fiber, so that the round-trip OPD within the cavity equals $2nL$, with n the refractive index of the fiber mode. As the position of the translation stage is scanned, an interference pattern will be observed when $|2nL - L_s| < L_c$, where L_s is the OPD of the SLD Michelson interferometer and L_c is the coherence length of the SLD. The maximum amplitude of the fringe pattern (center of the central fringe) is observed when $L_s = 2nL$. Fringe patterns from the sensing FFPI, the reference FFPI, and the DFB interferometer are collected by a data acquisition board and processed by a personal computer.

In 2010, an extrinsic FPI-based high-temperature pressure sensor was invented for avionics application [15]. On the other hand, the development of instruments for detecting is also important for universe exploration. A new tunable FPI spectral camera has been reported in 2011, which makes it possible to collect spectrometric image blocks with stereoscopic overlaps using lightweight unmanned airborne vehicles (UAVs) platforms. It is an important sensing application, because this technology is increasingly needed in various environmental measurement and monitoring applications.

The FPI-based spectral camera developed by the VTT provides a new way to collect spectrometric image blocks. The imager is based on the use of multiple orders of FPI together with the different spectral sensitivities of red, green, and blue pixels of the image sensor. With this arrangement, it is possible to capture three wavelength bands with a single exposure. When the FPI is placed in front of the sensor, the spectral sensitivity of each pixel is a function of the interferometer air gap. By changing the air gap, it is possible to acquire new set of wavelengths. With smaller air gaps, it is also possible to capture only one or two wavelengths in each image. Separate short-pass and long-pass filters are needed to cut out unwanted transmissions at unused orders of the FPI. During a flight, a predefined sequence of air-gap values is applied using the FPI camera to reconstruct the spectrum for each pixel in the image. The principles of the FPI spectral camera have been described by Saari et al. [16] and Mäkynen et al. [17].

The general data processing chain for FPI spectral image data has been shown in Figure 6.7. The processing chain includes data collection, FPI spectral data cube generation, image orientation, digital surface model (DSM) calculations, radiometric model calculations,

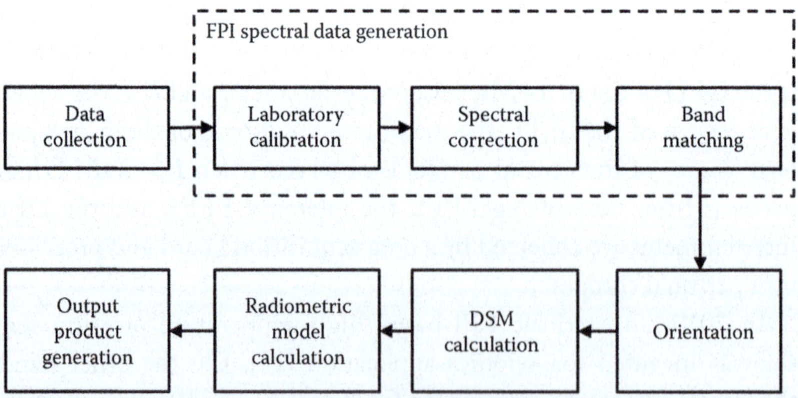

Figure 6.7 Data processing chain for FPI spectral image data.

and output product generation. When developing the data processing chain for the FPI spectral camera, our objective is to integrate the sensor-specific processing steps into our existing processing line based on commercially available photogrammetric and remote sensing software. The FPI spectral data cube is a crucial step in the data processing chain, which includes three major phases: radiometric image corrections based on laboratory calibrations, spectral smile corrections, and band matching. The challenging part of processing the FPI data is that the bands in the spectral data cube are collected with a small time delay. The approach mentioned above was to select a few reference bands and determine their exterior orientations. The FPI sensor provides many alternative ways for processing the data.

Besides, an FFPI was also used as an optical resonator in a dual-frequency optical source system for navigation application in 2015 [18]. In this system, an optical reference source provides output signals at frequencies, where output difference frequency is in a specific range.

6.4 Biomedicine

The majority of commercial sensors widely used in biomedicine are electrically active and hence not appropriate for use in biomedical applications, in particular, in high microwave/radiofrequency fields, ultrasound fields, or laser radiation associated with hyperthermia treatment, due to local heating of the sensor head and the surrounding tumor due to the presence of metallic conductors and corresponding electromagnetic interference, resulting in erroneous readings. Fiber-optic sensors can overcome these problems as they are virtually dielectric [19].

Today, most biomedicine or biochemical sensing takes place in laboratories. Tissue is invasively removed by biopsies and then analyzed *in vitro*. This is an invasive, time-consuming, and expensive process. Fiber-optic sensors have become a more attractive means because they can be integrated into hypodermic needles and catheters owing to their small size. Through the use of appropriate sensing method, they offer the capability to provide immediate *in vivo* monitoring of biochemical species and concentrations of interest in a minimally invasive manner.

Extrinsic fiber-optic sensor is an effective way for measuring biochemical and biomedicine parameters, in which some active material's properties are modulated by molecular concentrations of interest. An FP cavity will modulate the light spectrum incident upon it as a function of the reflectivity of its two parallel surfaces, their separation, and the indices of refraction inside and outside of the cavity.

A miniature fiber pressure sensor based on FP WLI principle with a desired diameter of 550 μm have been successfully developed by Pinet, Pham, and Rioux as shown in Figure 6.8 [20]. The sensor comprises a micromachined silicon diaphragm membrane adhered to a cup-shaped glass base as the pressure sensitive element. The FP cavity length varies as the environmental pressure changes. White light from a lamp is guided directly into the FP cavity, which modulates the signal with a low-coherence interference, thus coding the sensor cavity length.

Intra-aortic balloon pumping (IABP) therapy, which usually consists of inserting a catheter terminated by an inflatable balloon through the femoral artery, has been developed for over three decades. It has been one of the most efficient and popular technologies for life support especially when pharmacologic therapy fails or presents a high risk of mortality [21].

The experimental setup is presented in Figure 6.9 [20]. To test the sensor performance, a Bio-Tek Instruments' blood pressure system

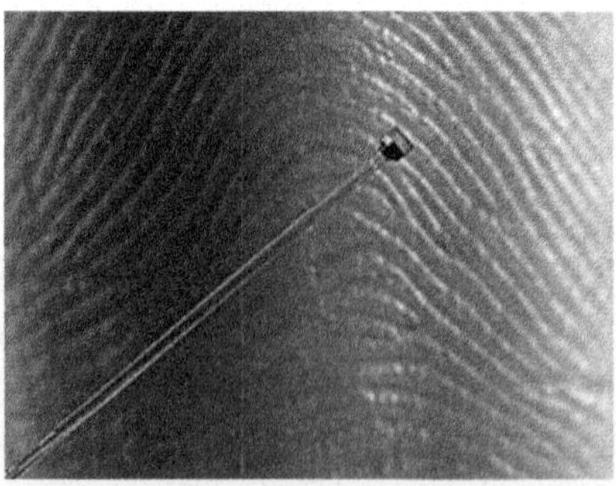

Figure 6.8 Miniature FP pressure sensor. (From Pinet, É., Pham, A., and Rioux, S. 2005. Miniature fiber optic pressure sensor for medical applications: An opportunity for intra-aortic balloon pumping (IABP) therapy. In Voet, M., Willsch, R., Ecke, W., Jones, J., and Culshaw, B., eds. *17th International Conference on Optical Fibre Sensors*, Bruges, Belgium. International Society for Optics and Photonics. vol. 5855, 234–237.

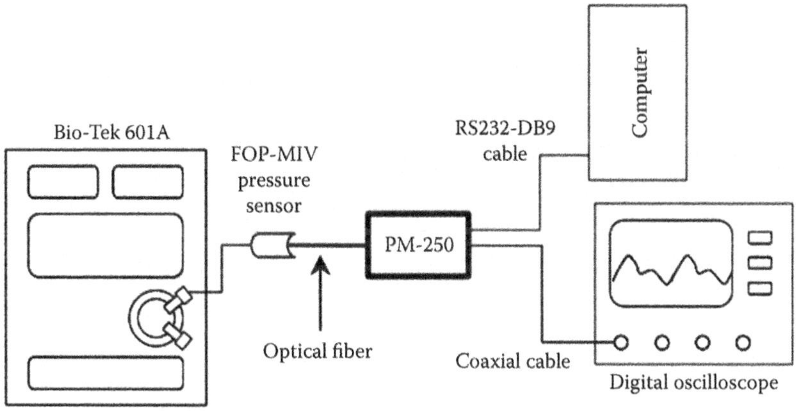

Figure 6.9 Experimental setup for pressure recording.

calibrator Model 601A was used as a calibrated waveform generator. This system simulates different waveforms including typical aortic pressure waves in a sealed liquid environment (water). The pressure was measured with the commercial FOP-MIV miniature pressure sensor connected to a commercial PM-250 signal conditioner operating at 250 Hz. A real-time signal was visualized on a digital oscilloscope. Data were transferred to a computer for further analysis.

The sensor is immune to electromagnetic interference that can occur in a hospital environment or during a surgery. Due to the white light cross-correlation technology used in the PM-250 signal conditioner, the pressure readings are not affected by the binding or displacement of the optical fiber, because the system measures the interference position directly related to the EFFPI cavity length of the pressure sensor and not a light intensity level, which would have been affected by optical fiber binding. The performance of the pressure sensor is shown in Figure 6.10.

Figure 6.10b shows that the FOP-MIV sensor is responding to square pressure waveform with high fidelity and without damping effects such as commonly observed in fluidic pressure transduction of commercial IABP systems.

EFFPI sensing in medical or biotechnology is also a useful tool. In 2006, an FP pressure chip for medical application was invented by Belleville and coworkers [22]. The chip has a body including a surface. A diaphragm covers the surface and is affixed to the body,

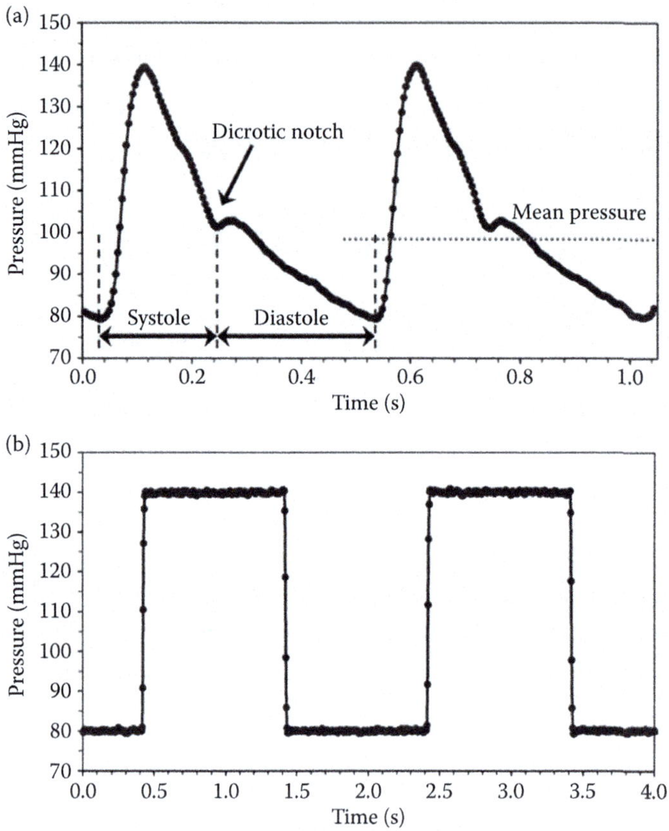

Figure 6.10 Pressure fluctuations simulated by Bio-Tek 601A pressure waveform generator as recorded using an FOP-MIV pressure sensor with a PM-250 conditioner. (a) Aorta tachycardia 120 bpm (80/140 mmHg). (b) Square pressure 30 bpm (80/140 mmHg).

where the diaphragm defines another surface. The surfaces are separated by a distance and form an FP cavity. The body has another cavity for receiving an extremity of an optical fiber. A small quantity of adhesive is required to secure the extremity of the optical fiber, where the body comprises a Pyrex portion. In 2007, Donlagic et al. developed a new manufacturing method of the FP optical sensor for medical application [23]. A cavity is formed by the shaped fiber end of a multimode step index optical fiber. A diaphragm arranged by a shaped optical fiber end forms a reflecting surface of a FP resonator. With the special manufacturing method, the sensitivity of the sensor is increased, and the production cost is reduced. A multi-cavity FPI sensor was invented in 2007 by Zhang and coworkers [24]. An FP cavity is formed at the end of the sensor to measure and provide temperature information. This device provides highly sensitive detection of the selected material, for example, deoxyribonucleic acid (DNA) hybridization, and bacteria, for the study of the thin-film and for chemical and biological material detection, even at low concentrations and without a need for labeling materials of interest, and prevents interference of the label with a process, for example, chemical reaction.

A simple and efficient approach for H_2-breath-test analysis based on FP microcavity has been reported in 2007 [25]. As shown in Figure 6.11, the microcavity is composed of a thin sensitive palladium layer exposed to hydrogen emission, and a GaP dielectric layer joined to a thick metallic mirror. The sensitive area consists only of the Pd film c. Insulating the substrate and the two films as indicated by the labels a and b, the optical properties of the sensor depend on the interaction between the exhaled hydrogen and layer c. In this configuration, layer a acts as a mirror. Its overall thickness is not significant above a certain value that provides maximum reflectivity and an essentially zero transmittivity. Reflectivity and transmittivity through this layer remain constant by increasing the thickness a above the

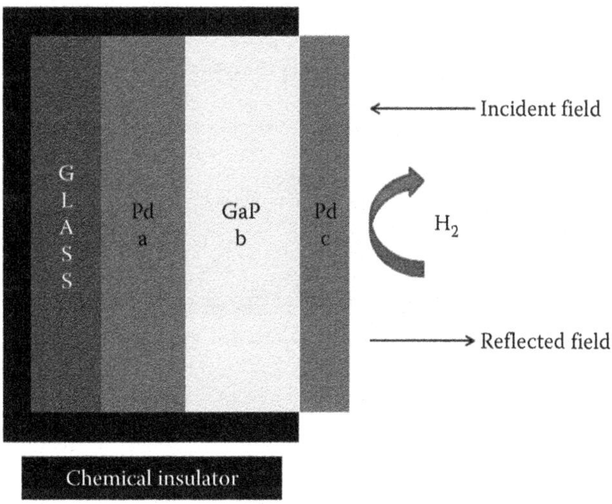

Figure 6.11 H_2-breath-test analysis based on FP microcavity.

aforementioned value. One might be tempted to think that the sensitivity of the device might increase by simply increasing the active layer c. However, the sensitivity is also related to the spectral position of the FP cavity resonance, which is uniquely determined by the thicknesses of the alternating layers b and c. The sensitivity strictly depends on the spectral position of the FP resonance, which in turn depends on the permittivity values resulting for different hydrogen concentrations. An extensive examination of the available geometrical parameters allows one to consider a wide number of resonances located at different spectral positions.

In 2011, Fan and coworkers demonstrated an FP cavity label-free biosensor with integrated flow-through micro-/nano-channel, which takes advantages of the large surface-to-volume ratio for analyte concentration and high detection sensitivity and built-in fluidic channels for rapid analyte delivery [26–28]. Recently, a novel EFFPI pressure sensor fabricated with a lensed fiber and a polymeric diaphragm was proposed for application in the medical field such as bladder or intracranial pressure measurement with reliable low-pressure measurement [29]. The structure is shown in Figure 6.12. A lensed SMF and a polymeric diaphragm form the cavity of the EFFPI. Two beams, which are reflected from the end face of the lensed fiber and the gold-coated surface of the polymeric diaphragm, respectively, are interfered through the SMF. Interference spectra of the pressure sensor were obtained while changing external pressure, and the cavity length was calculated by taking inverse fast fourier transform (IFFT) to the obtained interference spectra.

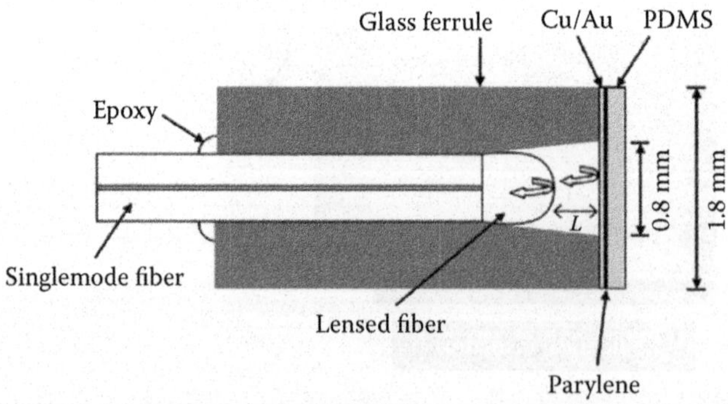

Figure 6.12 Structure of a pressure sensor.

So far, there have been commercial biomedical sensors based on the FFPI technology. FISO Technologies Inc. produces EFFPI sensors for clinical applications [20,30], which are really meaningful and useful for human health monitoring.

6.5 Other Applications

Besides the above applications, FFP sensors have also been used in other areas. A diaphragm-based EFFPI acoustic sensor was developed to measure attenuation properties of acoustic waves in water and a ~30° detection range was observed with an acoustic wave attenuation coefficient of 0.0626/cm [6]. An FP polymer film interferometer was constructed for ultrasound sensing that provides excellent detection sensitivities (<10 kPa) [7]. The development of an EFFPI ultrasound sensor offers the potential to use it in fiber-optic "smart structures" applications [9]. An inline FFPI fabricated from hollow-core PCF exhibits a wavelength pressure sensitivity twice as high as that of FBG, which is ~7.29×10^{-3} nm/MPa [31]. MEMS-based EFFPI sensors were first demonstrated by Mendez et al. for electric field sensing [32,33]. A compact EFFPI sensor that uses a wire-based magnetostrictive transducer [18] was investigated for DC magnetic field (100–35,000 nT) sensing [29]. Magnetic fluid is adopted to form an EFPI magnetic field sensor, which has a measurement sensitivity of 0.0431 nm/Gs [34] and 33 pm/Oe [35]. A nano-magnetic fluid-based EFFPI has also shown good results in the measurement of magnetic fields [24]. Others such as liquid-level sensors [36–38], flow velocity sensors [39], and humidity sensors [40,41], have also been used in their corresponding fields.

After many years of development, various FFPI sensors have been developed into commercial products. Typical FFPI sensors from different companies are listed in Table 6.1. Among them, the most successful FFPI sensor is one for pressure measurement in biomedicine, which is the best compromise offering, at affordable price, a great flexibility in terms of pressure ranges, high sensitivity, and miniature size suitable for most biomedicine applications, such as neuroscience, cardiovascular, intraocular pressure, urology, spine, bone, etc. Also, laser-micromachined FFPI sensors for measurement of physical parameters in harsh environments would have a very bright future [42].

Table 6.1 Typical FFPI Sensors from Different Companies

COMPANY	PRODUCT	TYPES OF SENSORS	SPECIFICATIONS	SAMPLES	APPLICATIONS
Opsens[a]	OPP-M	Pressure sensor	Pressure range: –50 mmHg to +300 mmHg Resolution: 0.2 mmHg		Medical
	OSP-A	Strain sensor	Strain range: ±5000 με Resolution: 0.5 με		Civil engineering and geotechnical applications
FISO[b]	FOP-MIV	Pressure sensor	Pressure range: –300 mmHg to 7500 mmHg Resolution: 2.6 mmHg		Medical
	Nortech TPT-62	Temperature sensor	Temperature Range: –40°C to 225°C Resolution: 0.1°C		Energy
Oxsensis[c]	PT1000™	Dynamic pressure sensor	Pressure range: 1–25 bar Dynamic output resolution: 2×10^{-5} of full scale Operating temperature range: –40°C to 750°C (continuous use) –40°C to 1000°C (R&D use)	–	Gas turbine acoustic instability monitoring applications
	TT1000™	Temperature sensor	Temperature measurement: Phase 1: From –55°C to 1090°C (–65°F to 2000°F) Phase 2: From –55°C to 1500°C (–65°F to 2750°F) Measurement update rate 1–10 Hz Resolution of 0.3°C (0.54°F)		Aircraft braking systems, gas turbine systems and furnaces.

(*Continued*)

Table 6.1 (***Continued***)

COMPANY	PRODUCT	TYPES OF SENSORS	SPECIFICATIONS	SAMPLES	APPLICATIONS
Prime photonics[d]	MFPS™ SP	Pressure sensor	Pressure ranges: 5–30,000 psi Temperature range: 0–750°F Frequency response: >100 kHz	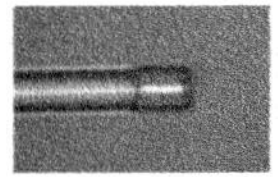	Medical, oil/gas, equipment monitoring
	MFPS™ ST	Temperature sensor	Temperature range: 0–1500°F Frequency response: 250 Hz	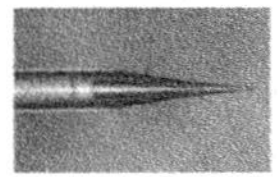	

[a] https://opsens.com/
[b] www.fiso.com
[c] http://www.oxsensis.com/
[d] http://www.primephotonics.com/

References

1. Farahi, F., Newson, T. P., Jones, J. D. C., and Jackson, D. A. 1988. Coherence multiplexing of remote fibre optic Fabry–Perot sensing system. *Optics Communications*, 65(5), 319–321.
2. Taylor, H. F. 2002. Fiber optic sensors based upon the Fabry–Perot interferometer. *In Fiber Optic Sensors*, Yin, S., and Yu, F. T. S., eds. New York: Marcel Dekker Inc. (pp. 41–71).
3. Lee, B. H., Kim, Y. H., Park, K. S., Eom, J. B., Kim, M. J., Rho, B. S., and Choi, H. Y. 2012. Interferometric fiber optic sensors. *Sensors*, 12(3), 2467–2486.
4. Gahan, D. F., Harpin, A. P. R., and Stevens, R. 2012. Optical sensor: U.S. Patent 8,253,945. 2012-8-28.
5. Donlagic, D., Cibula, E., and Pinet, É. 2013. Fabry–Perot optical sensor and method of manufacturing the same: U.S. Patent 8,559,770. 2013-10-15.
6. Needham, D. B. 1992. Multimedia interface device and method: U.S. Patent 5,167,021. 1992-11-24.
7. Donlagic, D., Pevec, S., and Cibula, E. 2014. Optical fiber sensors having long active lengths, systems, and methods: U.S. Patent 8,655,117. 2014-2-18.
8. Pevec, S. and Donlagic, D. 2011. All-fiber, long-active-length Fabry–Perot strain sensor. *Optics Express*, 19(16), 15641–15651.
9. Chen, M. W., Fu, Y., Zhu, Y., and Huang, S. 2004. Progress of optical sensor system for health monitoring of bridges at Chongqing University. *Proceedings of SPIE—The International Society for Optical Engineering*, 2360, 77–84.
10. Zhu, Y., Wang, X., Xu, J., and Wang, A. 2006. Optical fiber pressure and acceleration sensor fabricated on a fiber endface. U.S. Patent 7,054,011. 2006-5-30.
11. Liu, J., Sun, Y., and Fan, X. 2009. Highly versatile fiber-based optical Fabry–Pérot gas sensor. *Optics Express*, 17(4), 2731–2738.
12. Carisano, A. 2004. Minimization of temperature cross-sensitivity of EFPI pressure sensor for oil and gas exploration and production applications in well bores. *Proceedings of SPIE—The International Society for Optical Engineering*, 1(2), 206–212.
13. Wood, K., Brown, T., Rogowski, R., and Jensen, B. 2000. Fiber optic sensors for health monitoring of morphing airframes: I. Bragg grating strain and temperature sensor. *Smart Materials and Structures*, 9(2), 163–169.
14. Xie, Z. and Taylor, H. F. 2006. A fiber optic binary switch for aircraft application. In *2006 IEEE Sarnoff Symposium*, Princeton, NJ, USA (pp. 1–3).
15. Knobloch, A. J., Vernooy, D. W., Li, W., Shaddock, D. M., and Kennerly, S. J. 2011. High temperature optical pressure sensor and method of fabrication of the same, U.S. Patent No. 7,966,887. Washington, DC: U.S. Patent and Trademark Office.

16. Saari, H., Pellikka, I., Pesonen, L., Tuominen, S., Heikkilä, J., Holmlund, C. et al. 2011. Unmanned aerial vehicle (UAV) operated spectral camera system for forest and agriculture applications. *Proceedings of SPIE—The International Society for Optical Engineering*, 8174(1), 466–471.
17. Mäkynen, J., Saari, H., and Kai, O. 2011. Unmanned aerial vehicle (UAV) operated megapixel spectral camera. *Proceedings of SPIE—The International Society for Optical Engineering*, 8186(1), 81860Y1–81860Y9.
18. Vahala, K. and Li, J. 2015. Dual-frequency optical source. U.S. Patent Application No. 2015/0,236,789 A1. 2015-8-20.
19. Rao, Y. J., Deng, M., Duan, D. W., and Zhu, T. 2008. In-line fiber Fabry–Perot refractive-index tip sensor based on endlessly photonic crystal fiber. *Sensors and Actuators A: Physical*, 148(1), 33–38.
20. Pinet, É., Pham, A., and Rioux, S. 2005. Miniature fiber optic pressure sensor for medical applications: An opportunity for intra-aortic balloon pumping (IABP) therapy. In Voet, M., Willsch, R., Ecke, W., Jones, J., and Culshaw, B., eds. *17th International Conference on Optical Fibre Sensors*, Bruges, Belgium. International Society for Optics and Photonics. vol. 5855, pp. 234–237.
21. http://www.fiso.com
22. Van Neste, R., Belleville, C., Belleville, N., and Bussiere, S. 2006. Adhesive-assembled fiber-optic interferometer: U.S. Patent Application No. 2006/0,233,484 A1. 2006-10-19.
23. Pinet, É., Cibula, E., and Đonlagić, D. 2007. Ultra-miniature all-glass Fabry–Pérot pressure sensor manufactured at the tip of a multimode optical fiber. *Proceedings of the SPIE*, 6770(6770), 67700U–67700U-8.
24. Zhang, Y., Cooper, K. L., and Wang, A. 2007. Multi-cavity Fabry–Perot interferometric thin-film sensor with built-in temperature compensation: U.S. Patent Application No. 2007/0,115,480 A1. 2007-2-24.
25. Vincenti, M. A., De Sario, M., Petruzzelli, V., D'Orazio, A., Prudenzano, F., De Ceglia, D., and Scalora, M. 2007. Fabry–Perot microcavity sensor for H_2-breath-test analysis. *Journal of Applied Physics*, 102, 074501.
26. Guo, Y., Li, H., Reddy, K., Shelar, H. S., Nittoor, V. R., and Fan, X. 2011. Optofluidic Fabry–Pérot cavity biosensor with integrated flowthrough micro-/nanochannels. *Applied Physics Letters*, 98(4), 041104.
27. Fan, X. and Guo, Y. Optofluidic sensor used as stand-alone bio/chemical or physical sensors comprises cavity with reflective surface forming Fabry–Perot resonator and capillary with fluidic channel having a surface, where capillary positioned within the cavity. University of Michigan (Unmi-C), p. 63.
28. Fan, X. and White, I. M. 2011. Optofluidic microsystems for chemical and biological analysis. *Nature Photonics*, 5(10), 591–597.
29. Eom, J., Park, C. J., Lee, B. H., Lee, J. H., Kwon, I. B., and Chung, E. 2015. Fiber optic Fabry–Perot pressure sensor based on lensed fiber and polymeric diaphragm. *Sensors and Actuators A: Physical*, 225, 25–32.

30. Pinet, E. and Hamel, C. 2007. True challenges of disposable optical fiber sensors for clinical environment. In Third European Workshop on Optical Fibre Sensors, Napoli, Italy. International Society for Optics and Photonics. vol. 6619, pp. 66191Q-1–66191Q-4.
31. Rao, Y. J., Wang, W., Zhu, T., and Duan, D. W. 2009. In-line fiber-optic Fabry–Perot ultrasound sensor formed by hollow-core photonic-crystal fiber, IEEE Sensors 2009 Conference, Canterbury, New Zealand (pp. 858–860).
32. Mendez, A., Morse T. F., and Ramsey, K. A. 1993. Fiber optic electric-field microsensor, fiber optic and laser sensors X. Boston, MA. International Society for Optics and Photonics, 1795, 153–164.
33. Mendez, A., Morse, T. F., and Ramsey, K. A. 1993. Micromachined Fabry–Perot interferometer with corrugated silicon diaphragm for fiber optic sensing applications. Integrated optics and microstructures, Boston, MA. *International Society for Optics and Photonics. Proceedings of the SPIE*, 1793, 170–182.
34. Lv, R. Q., Zhao, Y., Wang, D., and Wang, Q. 2014. Magnetic fluid-filled optical fiber Fabry–Pérot sensor for magnetic field measurement. *IEEE Photonics Technology Letters*, 26(3), 217–219.
35. Zhao, Y., Lv, R. Q., Ying, Y., and Wang, Q. 2012. Hollow-core photonic crystal fiber Fabry–Perot sensor for magnetic field measurement based on magnetic fluid. *Optics and Laser Technology*, 44(4), 899–902.
36. Lü, T. and Yang, S. 2007. Extrinsic Fabry–Perot cavity optical fiber liquid-level sensor. *Applied Optics*, 46(18), 3682–3687.
37. Tong, B., Li, M., and Li, Y. 2012. Double-fiber Fabry–Perot Interferometry optical fiber liquid level sensor. In Asia Pacific Optical Sensors Conference, Sydney, Australia. International Society for Optics and Photonics. Proceedings of the SPIE, vol. 8351, pp. 83510T–83510T.
38. Lai, C. W., Lo, Y. L., Yu, J. P., and Chuang, C. H. 2012. Application of fiber Bragg grating level sensor and Fabry–Perot pressure sensor to simultaneous measurement of liquid level and specific gravity. *Sensors Journal, IEEE*, 12(4), 827–831.
39. Schlüßler, R., Fischer, A., Büttner, L., and Czarske, J. 2012. Application of a Fabry–Pérot interferometer for multi-point flow measurements. In Proceedings of the DGaO, Cologne, Germany, pp. 1–2.
40. Xu, W., Huang, W. B., Huang, X. G., and Yu, C. Y. 2013. A simple fiber-optic humidity sensor based on extrinsic Fabry–Perot cavity constructed by cellulose acetate butyrate film. *Optical Fiber Technology*, 19(6), 583–586.
41. Donlagic, D., Cibula, E., and Pinet, É. 2010. Single piece Fabry–Perot optical sensor and method of manufacturing the same: U.S. Patent 7,684,657. Washington, DC: U.S. Patent and Trademark Office.
42. Rao, Y. J. and Ran, Z. L. 2013. Optic fiber sensors fabricated by laser-micromachining, *Optical Fiber Technology*, 19(6), 808–821.

Index

F

S

V

W

Y

For Product Safety Concerns and Information please contact our EU representative GPSR@taylorandfrancis.com Taylor & Francis Verlag GmbH, Kaufingerstraße 24, 80331 München, Germany

Printed and bound by CPI Group (UK) Ltd, Croydon, CR0 4YY
14/09/2026
02216229-0004